BOILERS SIMPLIFIED

A GUIDE TO BASIC BOILER OPERATION

Lionel Edward LaRocque

BNP Business News Publishing Company
Troy, Michigan

Administrative Editor: Phillip R. Roman

Printed in United States of America

7 6 5 4 3 2 1

Library of Congress Cataloging in Publication Data

LaRocque, Lionel Edward, 1937—

Boilers simplified.

1. Boilers—Handbooks, manuals, etc. I. Title.
TJ263.5.L37 1987 621.1'94 87-685
ISBN 0-912524-40-5

CONTENTS

ACKNOWLEDGEMENT

The author wishes to express sincere appreciation to the National Association of Power Engineers for their valuable assistance.

PREFACE

Boilers Simplified was written as a reference not only for newcomers to the industry of steam generation but also for those already employed as boiler operators. This guide presents a comprehensive study of stationary engineering at both the low pressure and high pressure boiler levels.

Initially, this text begins by examining the fundamental principles of steam generation, progresses to a study of boiler designs and continues by reviewing important safety procedures and precautions and finally water treatment.

Lionel Edward LaRocque

DEDICATION

This book is dedicated to my wife, Carolyn Sue, for her assistance and support.

INTRODUCTION

WATER & STEAM BOILERS

HEATING SYSTEMS

Heating Mediums

Water: In a hot water or liquid system (commonly known as a hydronics system), water is the medium that is circulated through the system and acts as a vehicle to carry heat to its point of use. Basically, hot water is circulated from a boiler, through a piping system, to a heat exchanger such as a radiator or convector. This system utilizes a vacuum or booster pump to circulate hot water through a supply pipe to the heat exchanger. After passing through the heat exchanger, the water is then sent back to the boiler through the return pipe. In a closed loop system, the water is again reheated and is continually recirculated through the system.

Steam: A system utilizing vaporized water as its medium carries heat from a boiler (steam generator) through the piping

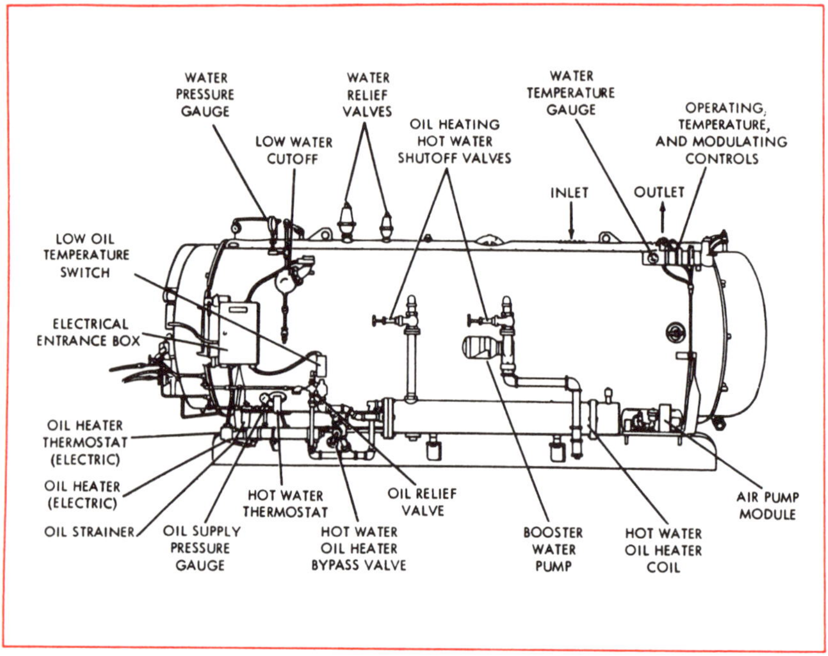

Figure 1. Typical hot water boiler. *(Courtesy Cleaver-Brooks)*

to the heat exchanger. But since steam has a greater volume than does water in its liquid state, pressure is created that, in turn, circulates the steam to the exchanger where the heat is extracted from the steam. The steam is then condensed into a liquid, and it is returned to the boiler where it is reheated. Once the water reaches its boiling point, it vaporizes and again becomes steam which is then recirculated through the system.

Figure 2. Hot water heating system configurations.

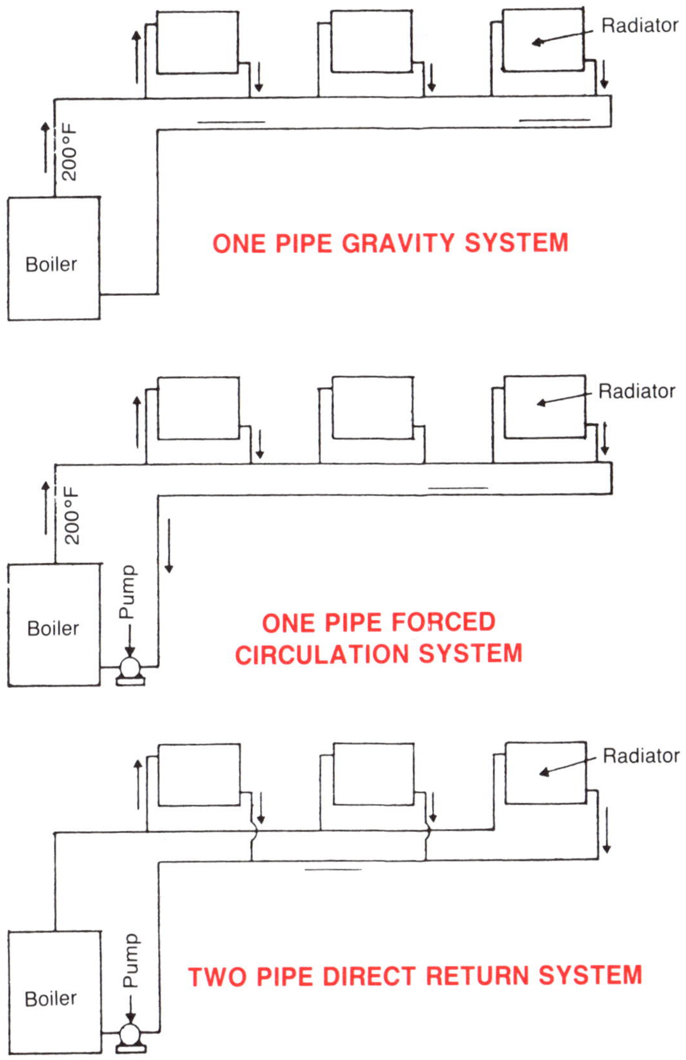

Other books available from
Business News Publishing Company

Hydronics
Good Piping Practice
Humidification Handbook
The Schematic Wiring Book Set
Refrigeration Licenses Unlimited
The Valve Selection and Service Guide
Marine Refrigeration and Fish Preservation
Systematic Commercial Refrigeration Service
How to Design Heating/Cooling Comfort Systems
How To Solve Your Refrigeration and A/C Service Problems
Getting Started in Heating and A/C Service
The (MSAC) Service Hot Line Handbooks
How To Make It In The Service Business
The A/C Cutter's Ready Reference
HVAC/R Reference Notebook Set
Industrial Refrigeration Service

For more information, write to:
Business News Publishing Company
Book Division, P.O. Box 2600
Troy, MI 48007

DISCLAIMER

1

HYDRONIC SYSTEMS

SECTION 1
HYDRONIC SYSTEMS

FUNDAMENTALS

A hydronic system uses water as the medium to transport heat from the point of generation to the point of use. Water has a greater specific heat than other fluids and, as a result, carries a greater amount of heat per pound. The term *heat energy* refers to the quantity of heat contained within a medium such as water. The amount of heat required to raise the temperature of one pound of water one degree Fahrenheit equals one Btu. Essentially, Btus measure the total heat content of a substance by its volume. Not only is water an excellent medium because of its capacity to carry heat, but it also has low viscosity and is non-corrosive to most materials.

In fact, the most economical way to move heat to a remote area is by piping in hot water. This is especially true when considerably long distances separate the points of use.

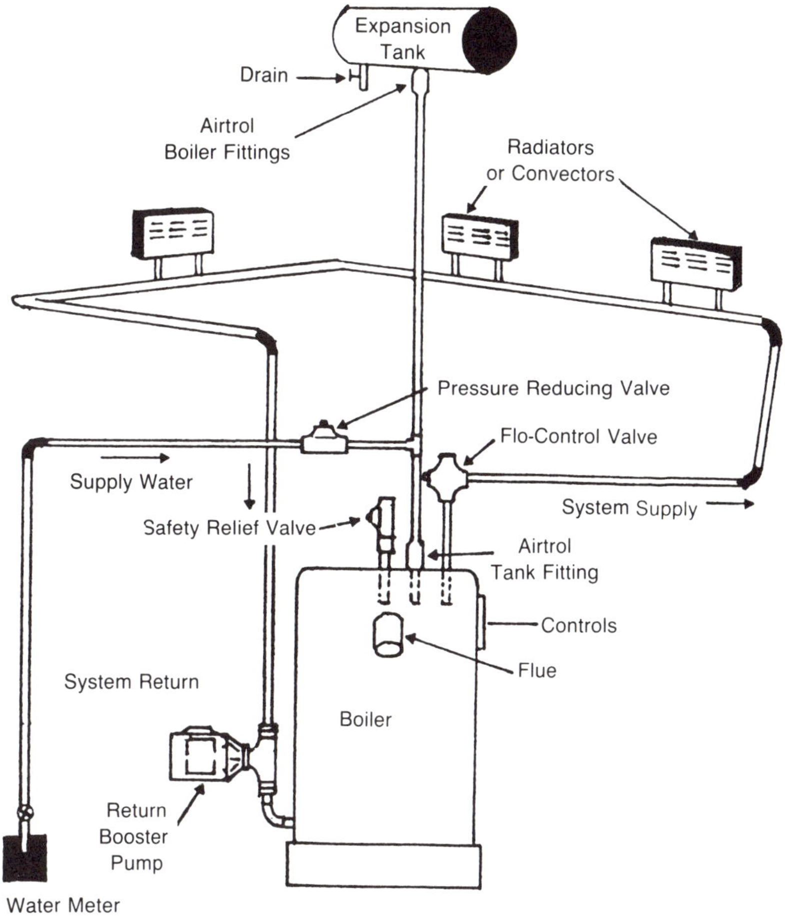

Figure 1-1. Basic hydronic system.

The primary components of a hydronic system are as follows:

- **Boiler** A boiler is a closed vessel. Constructed of cast iron or steel, the boiler contains and heats water for the system. Based on arrangement of tubes and the fuel or energy used, boilers are subdivided into two categories:

Firetube Boilers in which heat and combustion gases pass through tubes that are surrounded by water.

Watertube Boilers in which water passes through the tubes and the heat and combustion gases pass around the tubes.

- **Piping System** One or two pipe systems carry hot water to heat exchangers or terminals. (Refer to diagrams.)
- **Heat Exchanger** The common name for this device is either a radiator or convector. Basically, exchangers extract heat from water and transfer it into the area that requires heating.
- **Circulating Pump** The circulating pump cycles hot water through the piping system.

- **Boiler Accessories**

Expansion Tank As water heats up in a closed hydronic system, it expands. When the water expands, it compresses the air which is present in the system. Usually mounted above the boiler, expansion tanks installed in a system accommodate and control the compressed air displaced by the expanding water. Consequently, these units regulate or balance the pressure of the system.

When the system cools, water returns to its "unexpanded" state permitting the compressed air in the tank to re-expand to its normal pressure.

Pressure Relief Valve The most important component of the system is the pressure relief valve. If the system pressure exceeds the preset level of the relief valve (usually 15 psig), this safety device opens. Once open, the valve discharges water until the boiler pressure returns to a safe operating level.

Aquastat The aquastat is a sensory device. It maintains a constant supply of hot water in the boiler while the system is in the *standby* mode. The unit commonly consists of a thermostat bulb immersed in the water of the boiler. A pair of switches mounted on the outside of the boiler operates the device. The aquastat functions as both an operating control and a limit control. When the boiler water falls below the desired temperature, the aquastat opens an electrical circuit and activates the circulating pump. When the temperature of the boiler water exceeds or drops below the preset temperature range, the aquastat shuts down or activates the burner accordingly.

Electric Boiler This unit is basically a water tank equipped with insulated electric heating elements immersed in the water. The electronically controlled elements heat and maintain the water temperature.

STEAM BOILER BASICS

A steam boiler is a closed vessel, strongly constructed of steel or cast-iron, partly filled with water. Boilers heat water and convert it into steam for use in heating or as a power source. To heat water, boilers use heat sources that include solar energy, electricity and fossil fuels—gas, oil and/or coal.

To boil water and convert it into steam at atmospheric pressure, water must reach 212°F. One way a boiler can operate more efficiently is by applying heat to a larger surface area. This way the water can absorb more heat in a shorter amount of time. Installing a firetube boiler on its side, rather than installing it in an upright position, increases the heating surface (defined as any surface separating the fire and waterside of a boiler). Heating the inside of the boiler creates a combustion chamber. A combustion chamber is advantageous to a system. It makes fuel burn more efficiently by allowing the fuel and air to mix in greater quantities. The result, higher combustion temperatures.

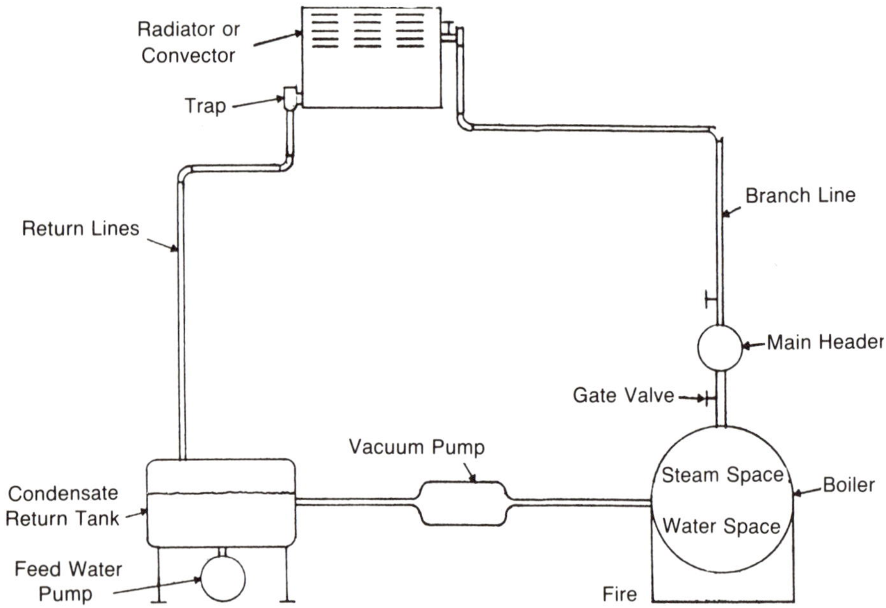

Figure 1-2. Steam to water cycle.

To further improve combustion, a chamber can be enlarged so the size of the heating surface increases. Subsequently, a grater amount of water will come in contact with the heated metal resulting in faster and more economical steam production.

FEEDWATER SYSTEM BASICS

As a boiler generates steam, the feedwater system must constantly replenish system water of the boiler. The feedwater system retrieves condensate, derrived from the steam at the heat exchanger, and returns it to the boiler for refeeding. A standby city-water line supplies the water necessary to make up the difference if there is not enough condensate to refill the boiler.

Steam leaves the boiler through a pipe and travels to a main header (a large diameter pipe where steam first accumulates). It then disperses into branch lines leading to the heat exchangers (radiators or convectors). At this point, the steam releases its heat to the radiator and condenses back into water (commonly called condensate).

A vacuum pump creates force a that helps draw the condensate out of the return lines and into a condensate tank. A trap, usually located near the exchanger, allows only condensate—and not steam, to travel back to the tank. The vacuum pump also circulates the water back to the boiler through pipes called feedwater lines. The boiler converts the returning condensate into steam again and recycles it.

FUEL SYSTEM BASICS

There are three common fuels used to heat water in a boiler: gas, fuel oils and coal. Each fuel requires its own system.

- **Gas System** The gas system burns either natural or bottle gas. Natural gas is an excellent fuel. It burns cleanly, costs less to use than manufactured bottle gas and requires no storage facility. In addition, natural gas needs no preparation prior to burning, as some other fuels do, lending it another advantage. Most boilers using natural gas (city-gas) have it piped to the boiler at the standard pressure. Because requirements differ according to boiler size and operating demands, the gas is not piped directly into the boiler at the existing line pressure. A regulator attached to the entry line reduces the gas pressure to zero pounds. The regulator then adjusts the gas pressure as necessary to satisfy the steam demand placed on the boiler. As the pressure-regulated gas enters the boiler, a blower supplies the correct amount of air necessary for combustion. The combined air/gas mixture then passes into the fire box or combustion chamber where a pilot flame of the burner ignites it. The flame produced at the heating surface brings the water to its boiling point and ultimately produces steam.
- **Fuel Oil System** Fuel oil systems call for special requirements due to the physical properties of the fuel it uses. Fuel oil is more stable than natural gas. Unlike natural gas, storage of sufficiently large quantities of fuel oil is necessary to provide a constant fuel supply to the boiler. Fuel oil is not highly volatile, so storage in either underground or aboveground tanks presents few safety concerns. A suction pump draws oil away from the tank and brings it to the boiler. At the boiler, oil leaves the pump under pressure. The increased pressure forces the oil through the fuel oil discharge line and on into the burner. More fuel oil enters the burner than is necessary for combustion. This extra fuel insures the presence of an adequate fuel supply so the boiler can meet its maximum steam demand. The excess fuel oil then returns to the tank by a return line.

- **Coal System** This fuel system utilizes two types of coal, anthracite (hard coal) and bituminous (soft coal). Coal feeding can be accomplished in two ways. One method is hand firing where coal is manually shoveled into a firebox. The second, stoker firing, is a mechanical means of feeding coal into the fire box. The two basic types of mechanical stokers are screw-feed stokers and ram-feed stokers (more commonly called under-feed stokers).

COMBUSTION BASICS

The air found in the earth's atmosphere contains many gaseous elements. Oxygen is one primary element that does more than just sustain life. Without oxygen, combustion can not take place. Consequently, to burn any fuel, air must enter the combustion chamber and mix with the fuel before the fuel ignites. Upon ignition of the air/fuel mixture, chemical changes occur.

During **complete combustion**, oxygen combines with the basic fuel elements (carbon and hydrogen) to form inert or harmless compounds, primarily carbon dioxide and water. When not enough oxygen is present, **incomplete combustion** occurs. If fuel does not burn completely, not only harmless end-products result. Soot, irritating aldehydes, sulfur dioxide and poisonous carbon monoxide result. To avoid these harmful end-products, a system needs to introduce an excess amount of air into the combustion chamber during the air/fuel mixing process. Although approximately only 50% of the air that enters the combustion chamber mixes with the fuel, the excess air supply helps ensure complete combustion.

DRAFT SYSTEM BASICS

When heated by the combustion process, air expands and rises. The warmer air becomes, the faster it rises or the greater its velocity. Following combustion, the hot flue (end) products rise and enter a large diameter pipe at the top of the boiler called the breeching. A fan blows the hot flue products out of the breeching, through exhaust vents and into a chimney or smoke stack. The hot flue products then rise up through the smoke stack and return to the atmosphere. The natural rising process of the hot flue products creates **draft**.

Draft occurs due to the difference between the outside or atmospheric pressure and the lower pressure existing inside the boiler system. This pressure difference induces an air flow through the system. The air moves with a velocity fast enough to overcome any resistance created by components within the boiler. The air flow provides an adequate supply of fresh air to ensure complete combustion. The draft created by the air flow also removes the flue products resulting from combustion.

A manometer is an instrument that measures draft. Both arms of this "U" shaped glass tube are open ended. When the tube is partially filled with water, both arms will have equal liquid levels. A flexible hose fastens over the end of one arm. The free end of the hose, attached to the breeching or stack, acts as a tap diverting the internal system draft to the manometer for measurement. If the manometer detects less than atmospheric pressure, the flexible hose draws water up that arm of the "U." The water level in the opposite arm will drop accordingly. If the instrument detects a level higher than atmospheric pressure, the water in the open side rises. If pressure in the test area is less than atmosphere, it is a negative reading. If the pressure is higher, it is a positive reading.

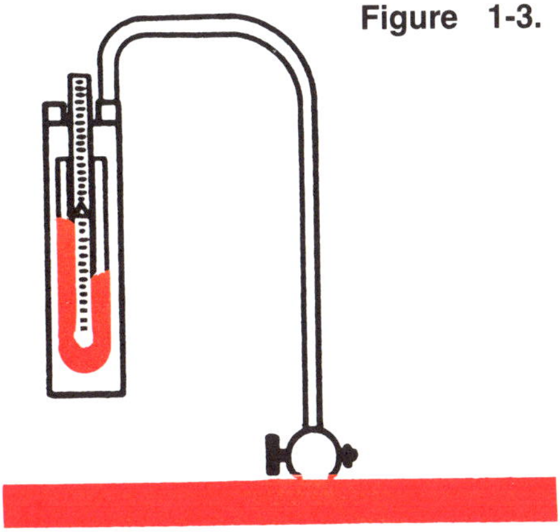

Figure 1-3. U-tube manometer.

Two types of draft exist, natural and mechanical. Natural draft occurs due to the difference in weight between a column of hot gas (in the chimney) and a column of cooler outside air of the same height. As the warm air in the chimney rises, the outside air moves in to replace it creating a draft. The rate of replacement depends upon the temperature difference between the two columns of air.

Increasing the height of a chimney can improve natural draft. A taller chimney increases the volume of warm air that the cooler air can displace.

Dampers or regulators can help control natural draft. These components can divert the air flow so it by-passes the combustion chamber and enters the chimney to dilute hot flue products. By cooling and diluting the flue gases with fresh air, the strength and velocity of the draft slows. Limiting the amount of air allowed to enter the combustion chamber and mix with fuel also controls the intensity of combustion. If the by-pass dampers or regulators close, the air must pass through the combustion chamber. By increasing the amount of fresh air entering the chamber, a greater quantity of air is available to mix with the fuel. The enriched air/fuel mixture promotes complete combustion which produces hotter flue products. As the temperature of the flue products entering the chimney increases, the faster the hot gases rise. In turn, the increased velocity creates a stronger draft.

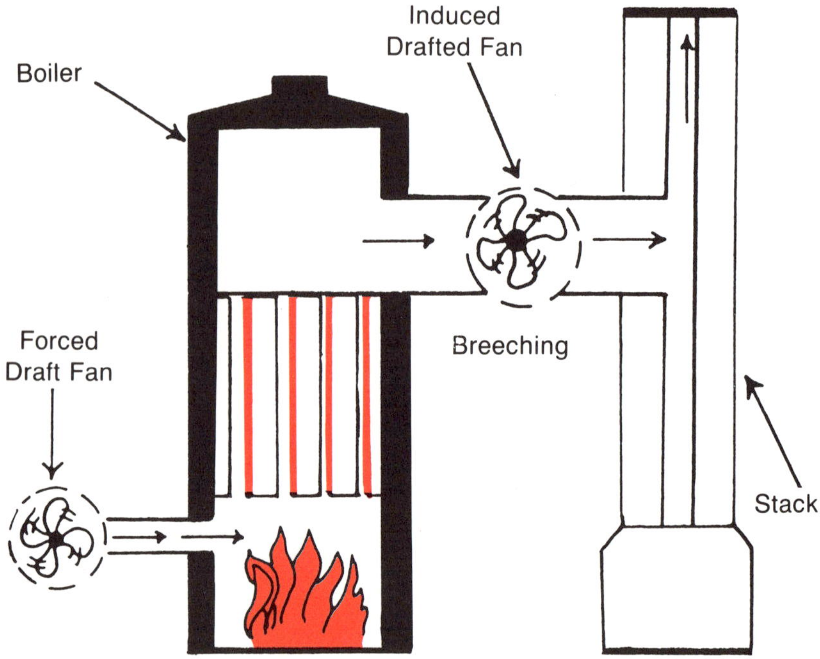

Figure 1-4. Forced and induced draft system.

Mechanical devices such as fans can also produce draft. Mechanical draft can either be *forced* or *induced*. In forced draft, a fan literally forces air into the fire box. As more air passes through the fuel, the fire burns with more intensity and becomes hotter. Under-feed stokers commonly use forced draft due to the considerable pressure required to force combustion air up through deep fuel beds. All large boilers that operate at high ratings usually come equipped with forced draft.

Rather than force air, induced draft creates a partial vacuum over the fire. The vacuum draws air up through the fuel bed to enhance combustion. An induced draft fan is usually mounted in the smoke pipe or breeching so the vacuum pulls air out of the boiler and pushes it into the chimney.

In either case, mechanical drafts are more easily controlled than natural drafts. Consequently, fuel burns more efficiently and produces more steam in systems equipped with mechanical draft. A balanced draft system combines automatically regulated forced and induced draft to keep the boiler furnace approximately at atmospheric pressure.

Automatic draft control regulates fans and dampers to increase or decrease air flow. This maintains constant steam pressure, as the load changes, and assures good combustion.

Devices that control mechanical draft often employ sensitive diaphragms. These detect changes in steam pressure or air flow. Once triggered, they initiate the opening or closing of fluid valves, electric switches and steam valves which control damper position or fan speed.

Unlike natural draft, changes in temperature or air flow do not affect mechanical draft. In addition, systems equipped with mechanical draft can burn inferior grades of fuel. Mechanical draft can also overcome resistance created by a tube bank when installing economizers between the boiler and chimney to utilize waste flue products.

HEAT TRANSFER BASICS

Heat produced by combustion in the boiler can be transferred to the system water or steam by radiation, convection or conduction. One or a combination of these three various modes may play a part in the transitional stages of heat transference within a boiler.

The direct transfer of heat in the form of radiant energy (liberated from burning fuel combustibles) is **radiation**. The amount of available heating surface which can be exposed to the heat of combustion determines the rate boiler's rate of radiant absorption.

Convection is the transfer of heat within a fluid due to particle movement or circulation that occurs. As hot particles in a fluid come in contact with heating surfaces, they continually replace cooler particles setting up a natural circulatory motion. Free convection is a condition that's strictly due to density disparity caused by temperature differences.

Heat transferred from one body to another without appreciable particle displacement is **conduction**.

Boiler surfaces heated directly by the flames of combustion on the fireside are considered the radiant surfaces. The remainder of the heating surface is considered the convection surface because only the hot gases of combustion, and not flames, come in contact with these surfaces. The amount of total heating surface and its distribution and the conductivity of the metal all affect the heat transfer capacity of a boiler.

Radiant surfaces receive the full radiant energy of the fuel and are subjected to the highest temperatures. The majority of the heat a boiler absorbs is derived directly from the radiant heat produced by the burning of fuel combustibles. Consequently, the radiant surfaces are of primary importance. It is necessary to keep fireside radiant surfaces free from soot and slag fouling. Fouling insulates radiant surfaces and hinders heat transfer which ultimately reduces overall boiler efficiency. Further, soot and other sulfuric fireside deposits have the potential of forming acids should moisture be present. The result, heating surface corrosion and eventual tube failure.

SECTION 1

FUNDAMENTALS

Review Questions

1. What term defines the science of heating and cooling with liquids?
2. Why is water considered a good heating medium?
3. What component moves and circulates hot water through the piping system?
4. Is a supply of fresh air required in this system considering its medium is water? Why?
5. What component senses water temperature and controls the burner cycle to maintain that temperature?
6. What is the function of an expansion tank?
7. List the three primary safety control units in a liquid heating system.
8. Name the most important safety device in any closed system system and describe how it operates.
9. How does an electric boiler heat water?
10. Define the term steam boiler.
11. List three sources of heat.
12. What is measured in Btus (British Thermal Units)?
13. What is the boiling point of water at atmospheric pressure?
14. How can steam boilers operate more efficiently?
15. Describe how a combustion chamber causes fuel to burn more efficiently.
16. Describe why increasing the heating surface of a boiler is is beneficial.

17. What is the basic function of the feedwater system?
18. What part of the steam return does the feedwater system primarily rely upon?
19. Describe the route which steam takes once it leaves the boiler.
20. What is a steam trap and how does it work?
21. How is condensate returned to the boiler?
22. List the three common fuels used for heating.
23. Name the two primary classifications of gas.
24. List three reasons why most systems use natural gas.
25. What reduces the city-gas line pressure before it enters a boiler?
26. Why must city-gas line pressure be reduced before the gas enters a boiler?
27. How is combustion air supplied?
28. Where does the air and gas mix in preparation for combustion?
29. What ignites the air/gas mixture?
30. What component supplies fuel oil to the boiler?
31. Trace the basic route of the fuel oil system.
32. What happens to excess fuel oil not consumed during combustion?
33. Identify the two most common types of coal (give both scientific and common names).
34. Define the function of a stoker.
35. List two common types of mechanical stokers.
36. Briefly define draft.
37. Why is draft necessary?
38. What are the two types of draft?
39. How can a stronger draft be created?
40. Can natural draft be controlled? How?
41. What type of draft acts independently from the effects of wind and temperature changes? Briefly state its benefit to a large scale system.

2 TYPES OF BOILERS

SECTION 2

TYPES OF BOILERS

There are many types of boilers and system configurations in use today. And although a wide array of boilers exists to meet specific design requirements, all operate in basically the same manner.

Simply stated, a boiler is an apparatus designed to convert fuel energy into a working medium (steam or hot water). The working medium, in turn, transfers heat. The heat transfer requirements greatly influence the boiler system size and set-up.

Heat source, fuel type, firing, shape and pressure are a few characteristics which distinguish one boiler from another. Primarily, though, the following factors determine boiler classifications.

- shell (drum) position: horizontal or vertical
- use: stationary, portable, tractor or marine
- furnace location: internally or externally fired
- tube position (relative to water and hot gases): firetube or watertube boilers

The two basic types of steel construction boilers are the firetube and watertube varieties.

FIRETUBE BOILERS

For the most part, firetube boilers have two principle uses, heating and producing industrial process steam. Firetube boilers lend these common applications decided advantages. First, firetube boilers can store a large quantity of water. The large water capacity of a firetube boiler enables it to diminish the harsh effects of abrupt steam demand fluctuations. Still, there is one draw-back to a large water capacity. The larger the water volume, the longer it takes to bring the boiler up to pressure from a cold start. A second advantage, firetube boilers are relatively inexpensive when compared to water-tube boilers.

Considered "shell boilers," firetube boilers contain combustion heat and gases within the confines of tubes surrounded by water. The firetube boilers include:

Scotch Marine: The Scotch Marine (SM) is the most widely used firetube boiler for both heating and industrial applications. The SM design features a short cylindrical outer shell with a wide diameter. The SM boiler contains one or more cylindrical furnaces in the lower half of the shell. After primary combustion, the hot gases pass into a brick lined combustion chamber at the back of the boiler. The gases then move up into the firetubes. These tubes return the gases to the front of the boiler. The gases next pass through large furnace flues [corrugated to withstand the stress of expansion and contraction] and into the breeching. Finally, the gases exit out the chimney.

Figure 2-1. The Scotch-Marine (SM) boiler is adaptable to the design tendency towards achieving a minimum of boiler heating surface with improved heat transfer techniques. The improvement in heat transfer is accomplished by designing for fluid fuels, controlling flue gas velocities and improving waterside circulation. The result—a compact high capacity boiler. *(Courtesy, Cleaver-Brooks)*

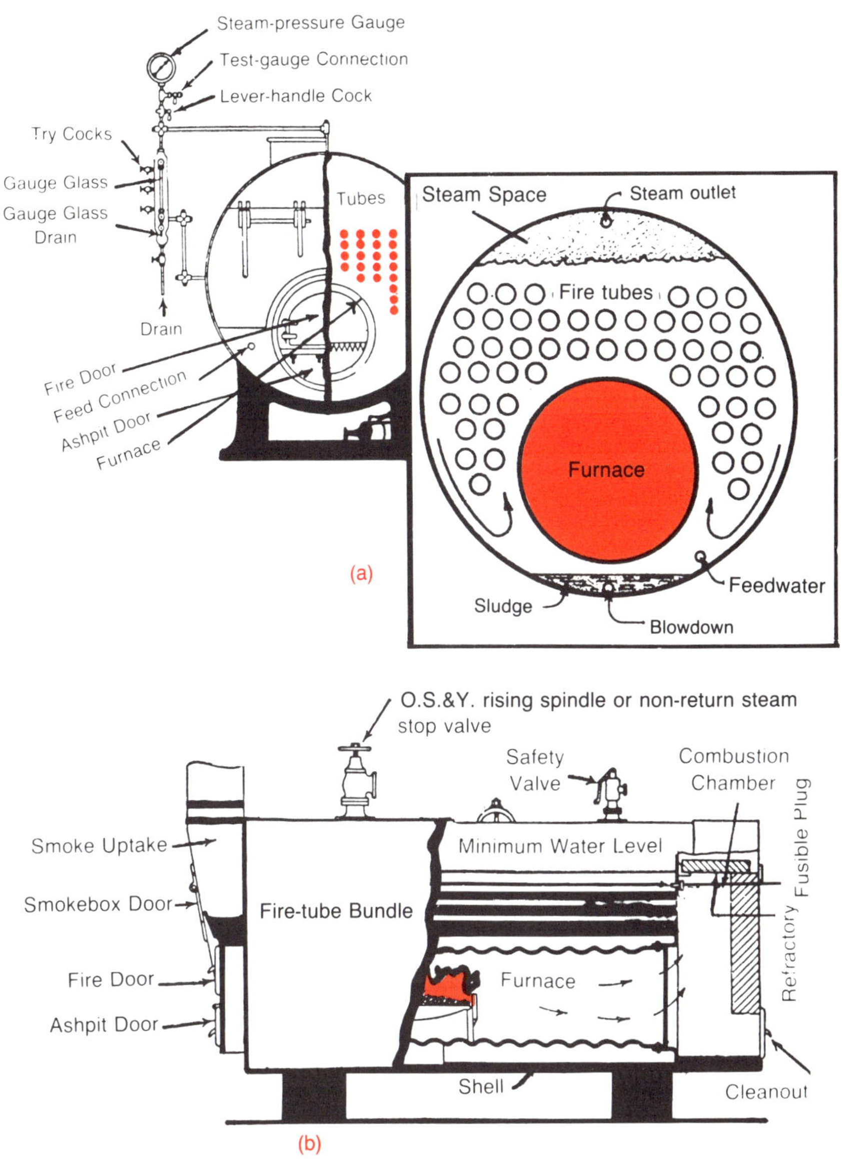

Figure 2-2. Dry-back Scotch Marine boiler: a) cross section views, b) longitudinal cut-away.

There are two types of SM designs. The **dry-back** SM boiler features a rear combustion chamber lined with a heat insulating (refractory) material—firebrick or fireclay. In comparison, if the rear combustion chamber has a water jacket, it is a **wet-back** SM boiler. In either case, the rear combustion chamber requires a water jacket or refractory lining to protect it against the intense heat produced by combustion.

Figure 2-3. Typical gas flow patterns in Scotch-Marine boilers.

Figure 2-4. SM boilers have a strong external resemblance. Internally, there are many important differences. The arrangement and utilization of the heating surface are far more crucial than the mere amount of surface. *(Courtesy, Cleaver-Brooks)*

Vertical Firetube Boiler: The Vertical Firetube Boiler is no longer used for industrial power installations. Since this boiler uses a minimum amount of floor space, it is still popular for home heating use. Heat from the basally located fire-box is transferred to the water through the fire-box plates. The hot gases resulting from combustion rise up through the firetubes then exit directly into the stack. As the hot gases pass through the vertical tubes enroute to the stack, they heat the water adjacent to the tubes. The hot water rises and forces the cool water down along the outer shell providing good circulation, the noteworthy characteristic of this boiler design.

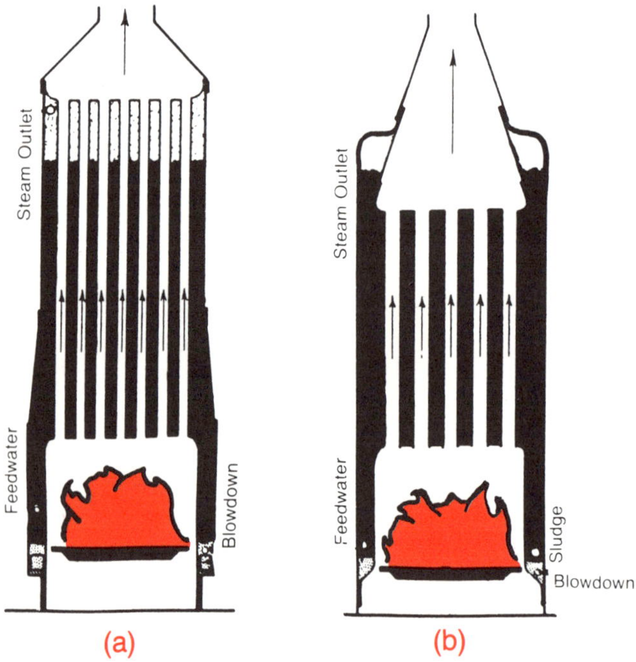

Figure 2-5. Vertical boiler sectional view showing gas flow: a) dry-top design, b) wet-top design (submerged head).

There are two types of these compact, self-contained internally-fired units: the **dry-top** and the **wet-top**. In both, an upper tube plate forms the top head of the boiler. In the dry-top, the minimum water level only reaches approximately two-thirds the height of the vertical tubes when measured from the lower (crown) tube sheet. The remaining portion above the water level contains steam.

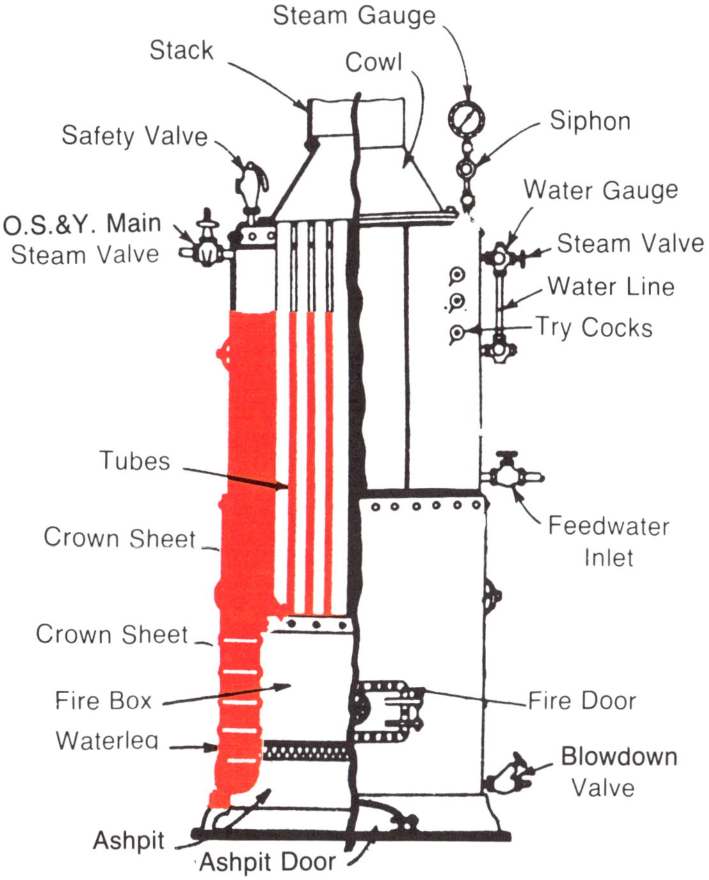

Figure 2-6. Cut-away of a dry-top vertical tube boiler.

Figure 2-7. Cut-away of wet-top design.

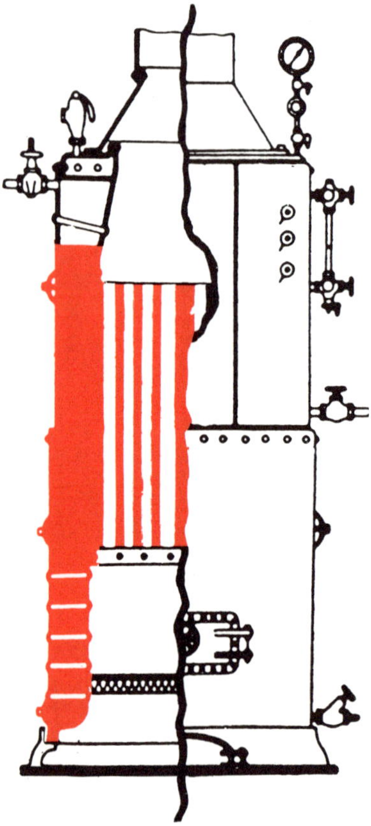

The wet-top design is similar to the dry-top except for the top head and operating water level. In the wet-top, a conical shaped smoke hood is attached to the upper tube sheet. This construction permits a larger water capacity by enabling the waterline to raise above the upper tube sheet.

Locomotive Boiler: Years ago, the Locomotive Boiler was a common sight. Not used for heating, these boilers functioned as portable power plants for tractors and locomotives. This boiler design even offered limited stationary use in oil drilling and sawmill operations.

The locomotive boiler is an internally fired, firetube unit. Mounted horizontally, this design consists of a cylindrical shell (barrel) attached to a furnace that has reinforced fire box water-jacket (water leg). The shell does not contain the furnace as is common in most firetube boilers. The furnace is usually rectangular and has a curved crown tube sheet. Large diameter firetubes, which permit minimal smoke stack heights, run the length of the shell to the front tube sheet at the *head* of barrel. Combustion gases pass through the fire-tubes, enter the smoke box (the space between the front tube sheet and barrel head) and rise through the smoke stack.

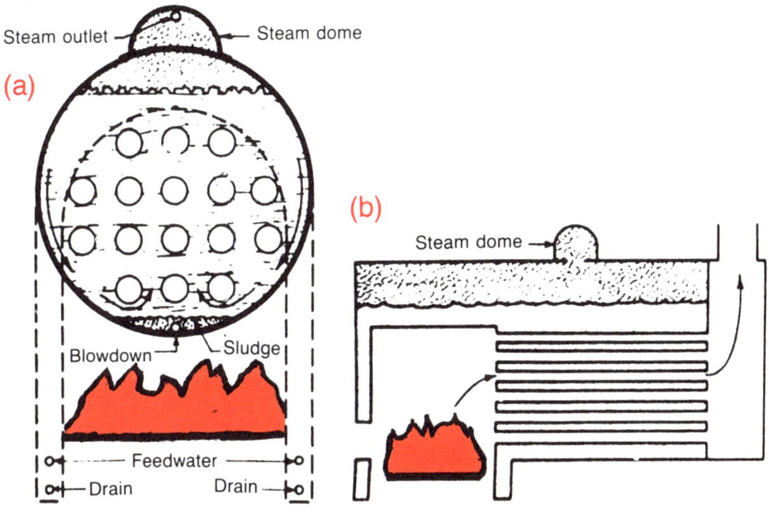

Figure 2-8. Sectional views of locomotive type boiler: a) cross section showing water circulation, b) longitudinal section showing gas flow pattern.

Locomotive boilers often have one or more steam domes on top of the barrel. These domes increase the space between the waterline and the main steam outlet. The dome height allows steam to be drawn off well above the water line. This reduces the risk of "priming" or having water carried over with the steam. Although this boiler design features large water capacity and steam release surface, circulation at start-up is sluggish due to the water volume.

Horizontal Return Tubular (HRT) Boiler: Probably the most common of the externally fired, firetube boilers, the HRT is a good, economical steamer. In the past, this boiler design occupied a secure niche as the primary source for medium and small industrial process steam applications. Due to the increasingly expensive labor required for initial installation, the HRT boiler has lost considerable popularity to the self-contained, internally fired Scotch Marine boilers.

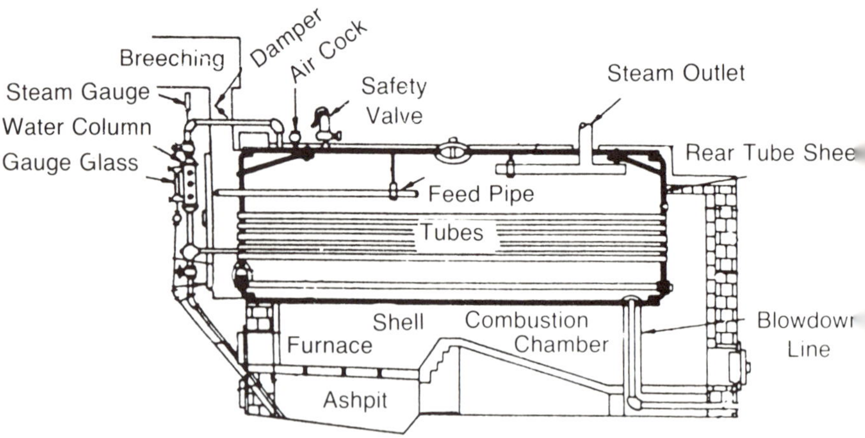

Figure 2-9. Horizontal-return tube (HRT) boiler showing main components.

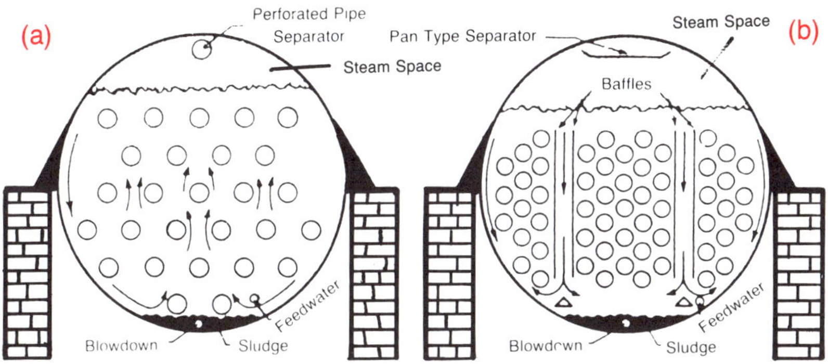

Figure 2-10. Cross sections displaying water circulation in HRT boiler: a) staggered tube arrangement, b) parallel tube arrangement.

An HRT boiler contains no fire box. Consequently, the boiler is set over an external brickset furnace firing. The brick setting (furnace/combustion chamber) is necessary for this design to provide combustion space. Grates incorporated into the brick setting beneath the front of the shell allow access to the combustion chamber for stoking.

The HRT cylindrical shell, capped with flat heads, contains a large cluster of 3 to 4-inch diameter firetubes running the length of the shell's internal waterspace. Inside the shell, steam collects above the waterline for separation and storage.

During operation, hot combustion gases from the furnace rise and sweep along the bottom of the shell towards the rear of the combustion chamber. The gases rise, pass through the horizontal firetubes and smoke box then discharge into the breeching. The gases then dissipate through the stack. The extended HRT return-tube arrangement provides a considerably longer exhaust path than does a direct tube arrangement.

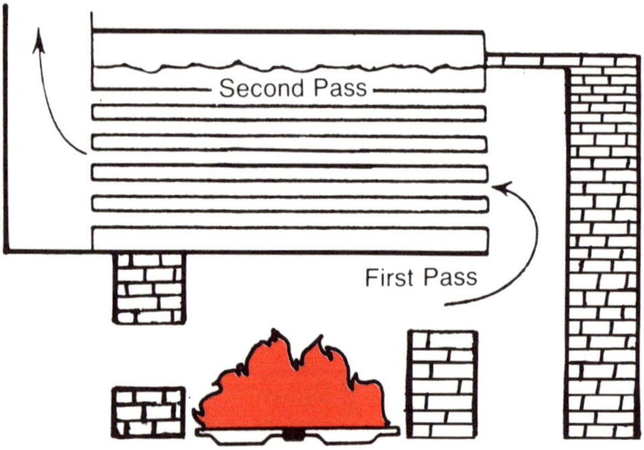

Figure 2-11. Sectional view showing gas flow pattern in HRT boiler.

And as the travel of hot combustion gases increases, the quantity of heat released by the gases increases proportionately. The additional heat absorbed by the system increases the quantity and rate of steam production.

Suspended by brackets fastened to the shell, small HRT boilers rest on the brickwork walls of the setting. Rather than suspend larger, heavier HRT boilers on the setting, bolts and lugs suspend them from over-head steel cross beams. Steel columns often lend additional support on larger installations.

Very few HRT boiler shells exceed 84-inches in diameter or exceed pressures of 200 lbs/in^2 (psi). Larger diameters or higher pressures structurally require shell plates thicker than ¾-inch. Since increased plate thickness slows down heat transference, thick plates exposed to intense heat tend to corrode because water cannot conduct the heat away from the metal surface quickly enough.

WATERTUBE BOILERS

A watertube boiler contains and circulates water **through** its tubes rather than around them as is the case with firetube boilers.

While the tubes of a firetube boiler are contained within the shell, watertube boilers typically have tubes arranged external to the shell. Since the shells (or drums) of watertube boilers often act as water storage vessels and do not need to contain tubular heating surfaces, the shells can be built with smaller diameters. The smaller shell diameters can consequently withstand higher pressures. The tubes on the outside of the shell also allow larger capacities that are not structurally permissible in firetube arrangements.

The external tube arrangement of watertube boilers lends added advantages. Watertube boilers are fast steamers due to the increased heating surface provided by the tubes. The intensified circulation that results also helps accommodate sudden fluctuations in steam demand. And because the tubes make up the heating surface, rather than the shell plates, the deteriorating effects of intense heat upon the shell plates becomes minimal.

Like the firetube boiler, many types of watertube boilers exist. Some of the different models are the straight-tube and bent-tube, single-drum and multi-drum.

STRAIGHT-TUBE BOILERS

Straight-tube boilers have banks of staggered tubes connected to headers positioned at a 5-15 degree incline to aid circulation. The headers (flat surfaces that facilitate tube connections) are connected to the drum via downtakes (supply water to the tube) and risers where steam and water are discharged from the high end of the tube bank. Positioned over the tube bundles, a single drum lies either longitudinally or across the axis of the tubes. The firebox is usually located beneath the high end of the tube bundle.

It is possible to modify the heating surface of a straight-tube boiler by changing a number of variables: tube length, number of tube rows, tube spacing, tube diameter and length. (Rarely does tube length exceed 18-feet.) By altering these tube parameters, pressures change.

This boiler is commonly used for low pressure applications and has become quite popular in the past few years due to the built in maintenance advantages. Briefly, straight-tube boilers can accommodate low head room, most components are easily accessible for inspection and cleaning, tubes are easily replaced and, finally, most have a low draft loss.

A few disadvantages also exist with this design. Steam and water separation at high steaming is often difficult. Steam production is hampered because of low circulation rates.

Circulation in the straight-tube design occurs when steam and heated water rise along the inclined tube cluster to the front header. The steam and water pass through the header into the riser travel up to the drum. Once in the drum, water circulates through the downtake to the rear header and moves back into the tubes completing the cycle.

Long Drum Boiler: The long drum boiler design features a drum positioned longitudinally over an inclined straight-tube bundle. Larger configurations may consist of two or more drums. The drums in long drum boilers are usually installed horizontally over the high-end tube header and not set on an incline over (or parallel to) the tube bundle.

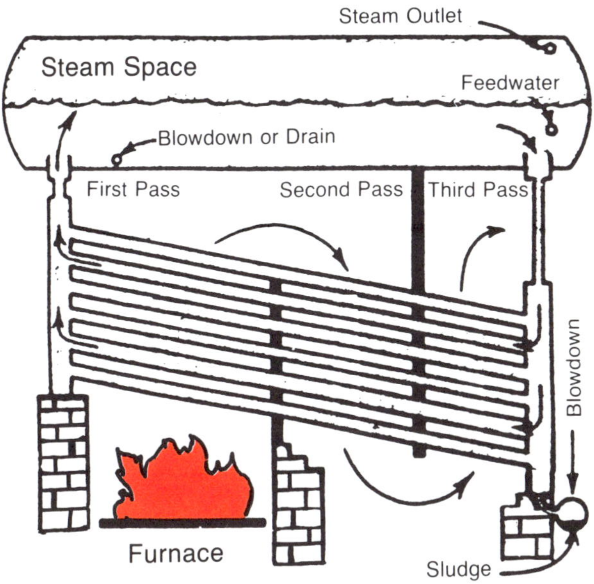

Figure 2-12. Water circulation in a long drum horizontal straight-tube boiler.

A mud drum located below the low-end header collects sediment. The mud drum facilitates periodic blowdowns to flush scale, rust and impurities from the boiler.

The tubes in long drum boilers are attached to the headers in such a way that their arrangement limits the number of sections that can be connected to a single drum. For example, a 48-inch drum can only accommodate a maximum of fourteen sections. If a boiler is designed to contain more than fourteen sections, a multiple drum configuration must be used. Due to this limitation, a long drum boiler is only suited for smaller applications where pressures are limited and where a large drum capacity is needed. One kind of small long drum boiler in common use today comes from Babcock & Wilcox manufacturing company.

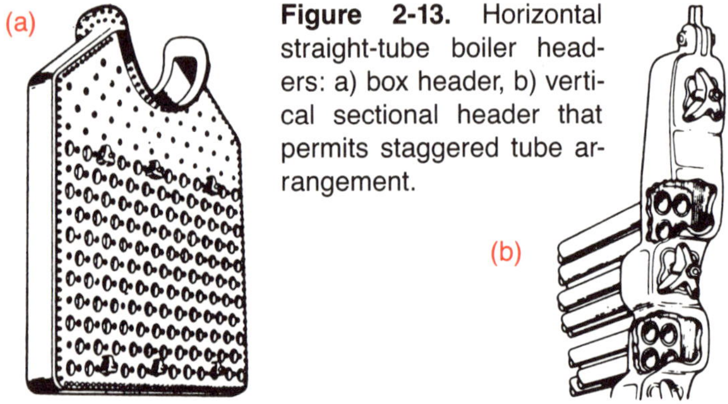

Figure 2-13. Horizontal straight-tube boiler headers: a) box header, b) vertical sectional header that permits staggered tube arrangement.

Cross Drum Boiler: The cross drum boiler design is configured so the drum sets crosswise over the low-end header of the tube bank. Positioning the drum transversely overcomes the limitations of the long drum boiler because the drum capacity can be increased by making the drum longer. The tubes are connected to the drum in longitudinal rows. And there is no limit on the number of tubes that can be connected to the drum.

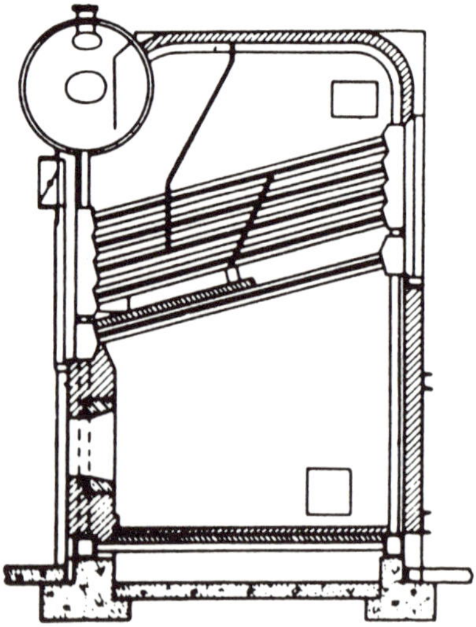

Figure 2-14. Cross drum horizontal straight-tube boiler.

By lengthening the drum, additional tubes can be connected increasing the boiler's capacity. Comparatively, long drum boilers require multiple tanks to increase capacity. Since only one tank is necessary in the cross drum boiler, the diameter of the drum can also be expanded to further increase its capacity.

The uniform discharge of steam into the entire length of a drum is the only way to obtain high capacity operation in a single drum boiler. One considerable draw back is then associated with any single drum boiler. As operating loads change, pressure may flucuate widely. This factor restricts the use of most single-drum boilers of this variety to small applications.

BENT-TUBE BOILERS

Bent-tube boilers offer distinct advantages over straight-tube designs. Bent-tubes are more flexible than straight-tubes and consequently yield:

- larger heating surfaces (allows higher steaming rates)
- excellent circulation
- drier steam production
- design flexibility (to suit space restrictions)
- easy access (inspection, cleaning and maintenance)

The advent of the bent-tube arrangement allowed boilers to be designed on a large scale with increased steaming capacities. Bent-tube boilers steam rapidly. They can respond to fluctuating steam demands quickly because only a relatively small amount of water is necessary to increase its steam generating capacity. Most of the larger power and central utility plant boilers fall into this boiler classification.

Essential and distinguishing components of this category of boilers are drums (or headers) and bent watertubes. The bent-tubes are arranged so that they actually form both the heating surface and the furnace enclosure. This tube enclosure (waterwall) functions as a water-cooling unit for the furnace and is also integrated with the heating or boiler surface (i.e., the two do not operate independently from each other).

By design, the drums of watertube boilers are positioned so they are protected from the direct radiant heat of the fire in the furnace. Because the dimensions of the bent-tube furnace can be modified to suit space restrictions and limitations, the capacity of the boiler can be increased without using larger drums.

It is also noteworthy to point out that if a tube ruptures, the water released will instantly flash to steam. In most instances, a serious explosion will not occur from a ruptured tube making watertube boilers somewhat safer than firetube boilers.

Stirling Watertube Boiler: Usually enclosed by a brick setting, this bent-tube boiler usually has three steam drums connected in series. The three drum series of the Stirling boiler hangs suspended from a frame of steel columns and cross beams. Short curved tubes connect the steam drums together to help balance pressure and water level. (Water level may vary somewhat during operation.) Suspended by tubes, an additional drum (mud drum) hangs below the steam drums to collect sediment and precipitates. The mud drum is suspended only by tubes. This method of suspension permits movement caused by contraction and expansion of boiler components due to temperature fluctuations that occur during final operation.

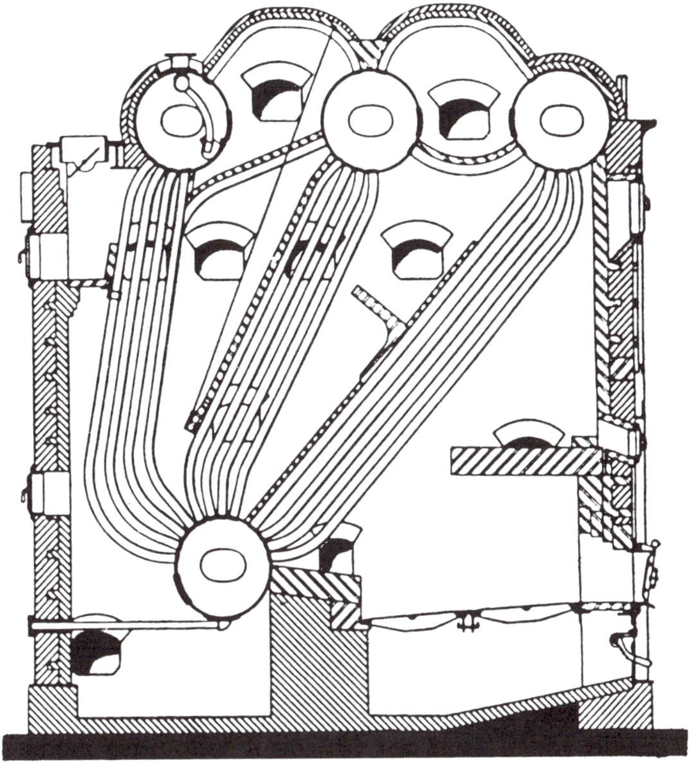

Figure 2-15. Stirling watertube boiler. *(Courtesy, Babcock-Wilcox Company)*

Circulation occurs as follows: feedwater enters the rear drum, passes down through the rear tube bank and then enters the front tube bank. Hot combustion gases deflected between the tubes by strategically positioned baffles heat the boiler's tube banks. As steam forms, it collects in the center steam drum.

Packaged Bent-tube Boilers: The bent-tube boilers easily lend themselves to the package boiler concept. Most major boiler manufacturers now offer various models of factory assembled and tested units. Mounted on structural steel bases, ready for installation on a simple foundation, one-piece units that incorporate firing equipment, mechanical draft systems, condensate return and feedwater makeup systems, controls and auxiliaries have nearly secured universal applications—except in instances where large central stationary units are necessary.

Today, bent-tube watertube boilers come shop assembled in three basic tube and drum configurations.

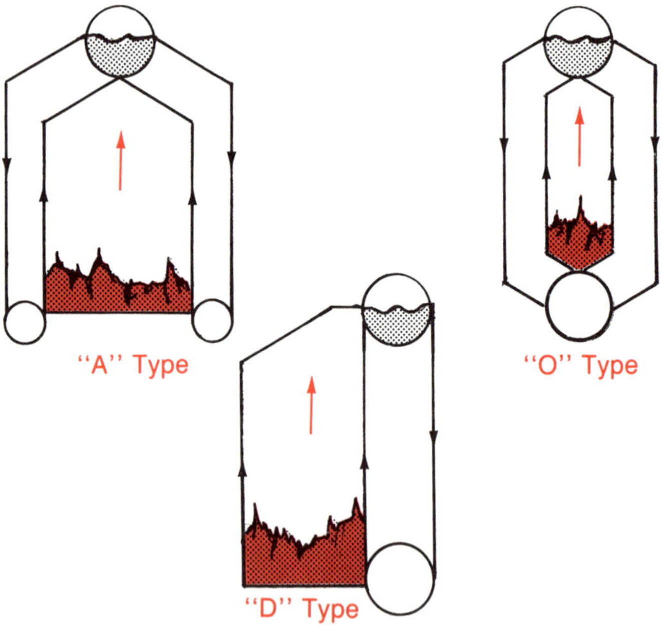

Figure 2-16. Common configurations of bent-tube watertube boilers.

A typical bent-tube packaged unit comes with a steel exterior casing, draft fan, burner, trim and controls all integral to the unit. The furnace is water-cooled, and most are suitable for pressurized operation although some can only operate at pressures under 250 psi. Some manufacturers offer larger units, equipped with sootblowers and superheaters, that operate under higher pressures which manufacturers stipulate in their operation guidelines.

The most prominent advantage attributed to packaged boilers is high combustion efficiency. Efficiency results in economic operation. Packaged units also save space, do not require elaborate foundations and many come equipped with burner systems that allow various fuels to be used for firing. Because they are portable, package units can be relocated and the majority of the equipment and materials salvaged.

One drawback of packaged units becomes evident when maintenance is required. The compactness of the units often makes servicing difficult. If the units are not operated according to the manufacturer's specifications, heating surfaces can quickly become coated with soot or scale. Coating of restricted surfaces can reduce operating efficiency and, ultimately, dependability.

CAST IRON BOILERS

Cast Iron (CI) Sectional Boiler: The cast iron boiler falls into the category of watertube boilers because water circulates inside its cast sections. Combustion gases remain outside the sections. Each section theoretically comprises a separate boiler.

Two types of ci boilers are commonly in use for commercial heating and to produce hot water, the vertical sectional and the horizontal sectional. In both, the heating surface is large in comparison to water volume, a trait that makes them relatively quick steamers. The vertical sectional ci boiler sections resemble pork chops and are assembled front to back. The assembled sections of a horizontal sectional are stacked like slices of bread over a flat base. Instead of being joined by tubes, the cast sections are connected by means of push and/or screw nipples or external headers.

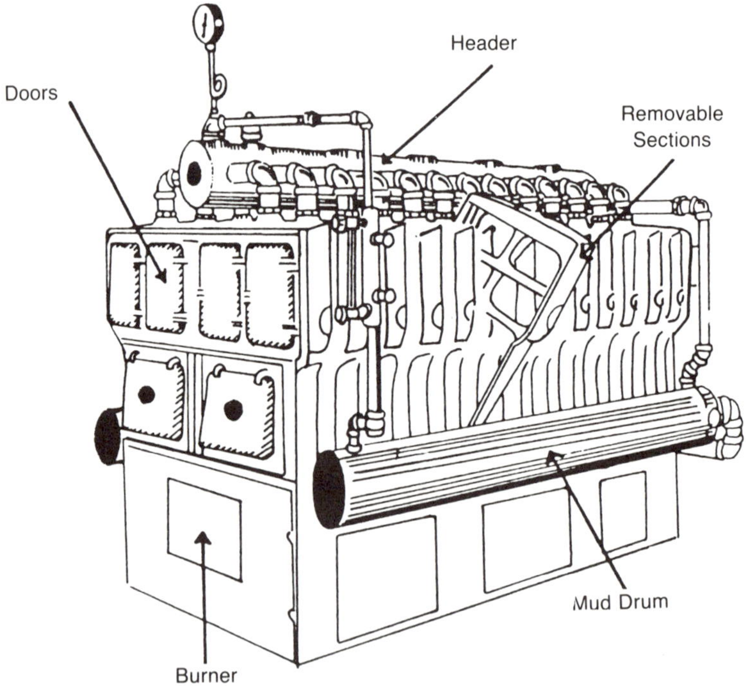

Figure 2-17. Typical cast iron sectional boiler.

The sections of the internal push nipple design are connected with tapered nipples and pulled together by tie-rods or bolts. Repair or replacement of a section often necessitates entire boiler dismantling.

With the external header boiler, the sections are connected to the header drums individually by screwed nipples. Usually, there is a steam drum at the top center of the boiler and a pair of return drums at the bottom. Each section in this arrangement can be replaced independently. The lack of cross connecting water passages produces a constant equalizing of the waterlines due to the positioning of the return drums. Even though overfiring this external header arrangement may crack adjoining sections, a dangerous explosion is not likely to occur.

Cast iron is an iron-carbon alloy that can molded into sections when in the molten state. When cast and allowed to cool slowly, a crust of graphite covers the casting giving it a grayish color. By nature, cast iron is brittle. The graphite coating that forms on a casting helps make it less brittle and more machineable.

Cast iron is quite resistant to corrosion. Still, scale and rust from untreated water can form and build up inside the ci sections restricting water circulation. This happens much in the same way sludge build-up restricts the circulation of motor coolant in an automobile. As water passages become fouled with mud or sludge, circulation gradually becomes more restricted. Not only does the system fluid pressure increase, but the the radiator can no longer sufficiently cool the fluid. Overheating then occurs. The build-up in ci sections eventually restricts water circulation to the point where it causes overheating.

If controls are not operating correctly or safety relief valves are not functional, overheating and increased pressures can cause a cast iron boiler to explode. When a ci boiler explodes, the brittle cast iron becomes fragmented and as lethal as shrapnel.

There are a number of reasons behind ci boiler explosions. And it has been noted that ci boilers will never fail twice for the same reason. Because of this, cast iron boilers can only operate under a maximum steam pressure of 15 psi (a limit established by the ASME Boiler Code).

Despite the threat of explosions, well maintained ci boilers offer excellent reliability, low maintenance cost and good longevity. An added advantage is that almost any kind of automatic fuel burning equipment can be installed and used efficiently in a ci boiler. And large combustion chambers and flue passages prevent draft loss allowing reduction of stack heights.

SECTION 2

TYPES OF BOILERS

Review Questions

1. What is a boiler?
2. Describe the primary difference between a firetube and watertube boiler.
3. Comparing the watertube and firetube boilers, which has the largest heating surface in the tubes?
4. What internally fired boiler often has more than one furnace? What advantage does a multiple furnace set up offer?
5. Can firetubes explode? Explain why or why not?
6. Which is supposedly safer, the watertube or firetube boiler Why?
7. What boiler is no longer used in large installations, but is still popular for heating homes?
8. What is the difference between a dry top and a wet top boiler?
9. What boiler is used for portable and steam tractors?
10. What is the function of a steam dome?
11. What watertube boiler is used mainly for low pressure applications?
12. What advantage does the cross drum boiler offer over the long drum boiler?
13. What are the three most common configurations of shop assembled watertube boilers?

14. Bent-tube boilers can have two, three or four drums. What What is the advantage of a multiple drum arrangement?
15. Why are equalizing tubes important in bent-tube boilers?
16. Why are circulating tubes bent in Stirling boilers?
17. What is the function of a mud drum? Why is a mud drum necessary for efficient operation?
18. Identify the two types of Scotch Marine boilers. Describe how the two designs compare and differ.
19. What characteristics constitute a good steam boiler?

3 SYSTEM COMPONENTS

SECTION 3

SYSTEM COMPONENTS

Due to the variety of boilers available in today's market, selecting boiler room components, which are compatible with the system design, is difficult. This fact accounts for the wide popularity of packaged boilers. Still, several different boilers of various designs may be equally well suited to meet the anticipated service conditions of a given installation. Even so, space restrictions such as low headroom or limited accessibility more often than not determine which boiler system is best to employ.

Despite the variety of available boiler designs, every component in a boiler system must meet the fundamental requirements of a given installation. If every component does not conform to those requirements, the system may not function correctly from the onset.

Whatever the size of the boiler, coordinating the fuel burning and heat generating equipment is essential to delivering steam or hot water at the required rate and condition (i.e., pressure, temperature and quality). For economical operation, heat has to be generated and delivered with minimal losses.

By today's standards, it is not enough that every component of the boiler be efficient and constructed well. The operating costs involved in steam or hot water production mandates that maximum over-all efficiency and reliability be achieved at minimal cost. To accomplish this, the boiler and all its component parts and auxiliaries must comprise a balanced design. Each element must be proportioned correctly and have proper relationship to every other element in the system.

FEED & FEEDWATER APPARATUSES

It is imperative boiler operators make sure the water in on-line boilers is maintained at the proper operating level to prevent damage due to overheating. When entering a boiler room, the operator's primary duty is to check and verify that the water level in all boilers is correct. If the water level is low, operators must know how to replenish the water supply.

An understanding of the basic water conversion cycle that occurs within a steam generating system (as well as the components involved) is necessary to operate boilers proficiently. The following is a description of the water/steam/water conversion cycle.

Water is fed into a boiler and heated until it turns to steam. The generated steam travels from the boiler steam space to the main header. From the main header it enters the main steam lines where it is diverted into smaller branch lines. The branch lines carry the steam to risers which channel the steam up to various heat exchangers, convectors or radiators

Figure 3-1. Packaged feedwater system for pressure applications up to 300 psi with intermittent operation. *(Courtesy, Cleaver-Brooks)*

where the steam gives up its heat. After moving through the heating units, the steam condenses back into water (condensate) and passes through a steam trap. This device traps the condensate and only allows steam to enter the steam return lines. The condensate which accumulates in the traps is either discharged as wastewater or routed back to a condensate tank through its own return line. Condensate stored in the tank is finally fed back into the boiler by a vacuum pump for recycling.

The feedwater line should always be piped into the boiler at a point just above the minimum safe water level. If for any reason the quantity of condensate returned to the boiler is insufficient to maintain a safe water level, a city water line must supply the additional or makeup water required to maintain the proper water level. Code stipulates that a stop valve be installed on the feedwater line where it enters the boiler. The stop valve is a safety precaution that permits the feedwater line to be shut off during any emergency.

As an added precaution, a swing check valve installed in the line prevents water in an overfilled boiler from backing up into the feedwater lines. The check valve is a hinged, one-way valve constructed so feedwater can pass and enter the boiler without obstruction. If the pressure in the feedwater line drops below the boiler pressure, the check valve closes and prevents steam or boiler water from backing-up into the feedwater line.

At certain times the boiler pressure may very well exceed the pressure of the feedwater line and hold the check valve closed. When the check valve is closed, condensate accumulates in the tank and must be emptied back into the boiler to make room for the new condensate constantly being generated. To overcome the pressure of the boiler and open the check valve, a vacuum pump is employed. When pumping condensate back to the boiler, the vacuum pump builds-up

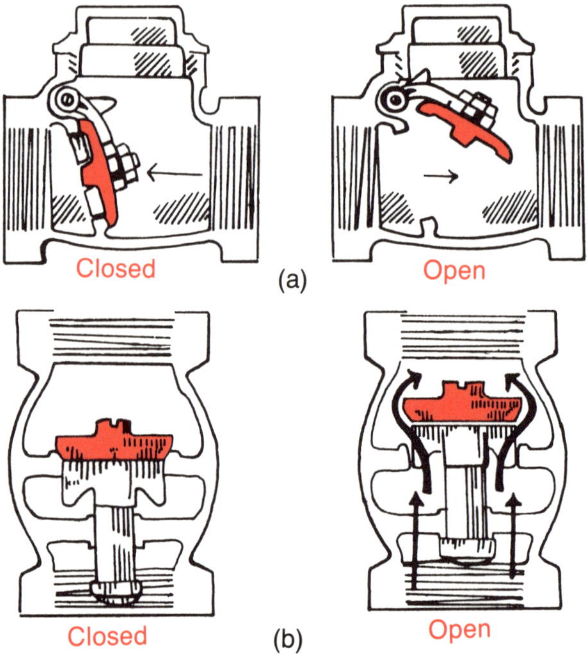

Figure 3-2. Check valves showing flow patterns: a) swing type check valve, b) vertical check valve.

sufficient pressure in the feedwater line to overcome the boiler pressure keeping the valve closed. The check valve then opens and the condensate is pumped back into the boiler for recycling. When the pump stops, the pressure from the boiler becomes dominant again and closes the valve.

The feedwater stop valve is usually a globe valve that is controlled manually. Globe valves are used in this capacity because they offer a restricted or Z shaped water passage that affords quick positive shut off. The globe valve design features a metal disk and seat that are easily replaced to facilitate repairs.

Feedwater Regulator: This device is designed to maintain constant water level in the boiler. Installed at the normal operating water level, the regulator is equipped with a sensor that automatically transmits a signal to activate the feedwater pump if the water level drops. Located directly under the condensate tank, the pump delivers enough makeup water from the tank to bring the water back up to the normal level. Once the water supply has been replenished, the sensor shuts off the pump.

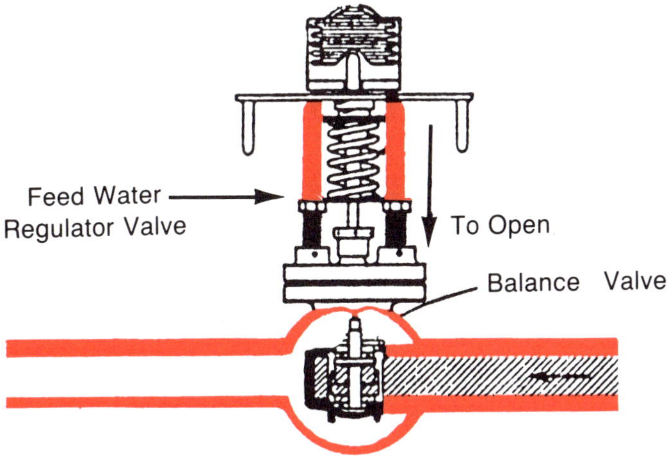

Figure 3-3. Feedwater regulator.

Makeup water is stored in the condensate tank which is fed by a vacuum pump. An automatic or manually operated city water supply line is also connected to the tank. If the vacuum pump cannot supply an adequate amount of condensate required for low water conditions from the return lines, the condensate is supplemented from the city water supply.

The automatic city water makeup system employs a float valve which is installed on the boiler slightly below the normal water operating level. As the water line drops, the float drops and opens the line to the city water supply. City makeup water must be piped into the condensate tank, rather than directly into the boiler, along with the condensate where the mixture can be chemically treated to remove impurities that damage heating surfaces and impair heat transfer before the mixture is pumped into the boiler.

Low Water Cut-Off: This device operates independently from the programming sequence control. If the water level drops to a dangerously low level, a low water cut-off shuts down a boiler by breaking an electrical circuit that deactivates the burner. An operator may encounter three basic types of cut-offs:

- The **float magnet cut-off** features a float rod with a ferrous plunger attached to one end. This rod/plunger assembly is encased by a nonferrous sleeve which shields the assembly from the field of a permanent magnet when the water level is normal. A mercury switch affixed to the magnet keeps the burner circuit closed. When the water level drops, the float rod assembly drops and reveals the ferrous plunger. When the small magnet pendulums towards the plunger, one end of the switch becomes elevated and opens the burner circuit.
- The **float linkage design** has a float attached, through a linked assembly, to a plate that supports a mercury switch. At normal water level, the switch keeps the burner circuit closed. When the water drops the float, the linkage tilts the plate and the switch opens the burner circuit.

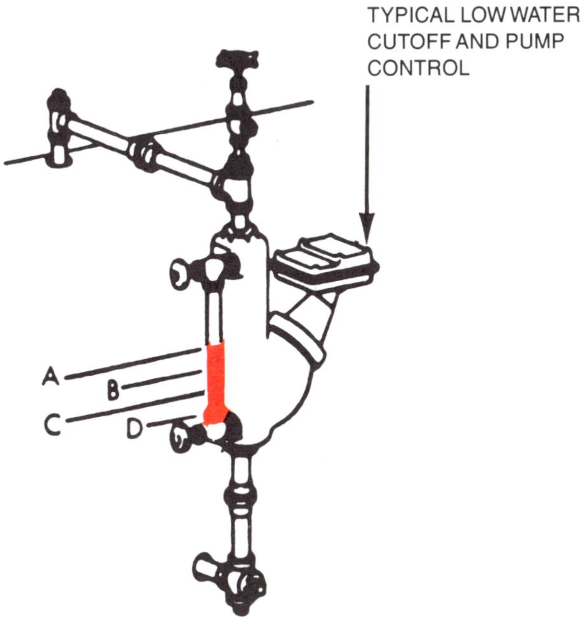

A. High level of water: Feed pump turns off at this point. Fill pressure vessel initially to this height.

B. Pump turns on when water level reaches B. Distance A-B is approximately 3/4 inch.

C. Low Water Cutoff Point: Burner will shut off if water level lowers to this point.

D. First visible point of gauge glass.

Figure 3-4. Typical low water cut-off.

- The **electrode probe design** utilizes boiler water to complete an electrical burner circuit. Usually positioned approximately 6-inches below the water glass (and always above the boiler fire box), this devices features an insulated electrode whose tip is kept partially underwater at the normal operating level. Current flows from the electrode through the water to the housing of the probe to maintain a closed circuit. When the water level drops and the electrode tip is no longer submerged, the circuit breaks to automatically shut down the burner.

Since the majority of boiler related accidents that occur are attributed to low water levels, the low water cut-off should be tested daily. Boilers are not exempt from serious damage such as overheating, implosions and explosions, despite the fact that most modern boilers are equipped with automatic devices.

Automatic devices often lend boiler operators a false sense of security which can foster disastrous results. Automated devices such as low water cutoffs may operate flawlessly for a given period of time, but eventually, they will fail. Any automatic device will fail. It is merely a matter of time. However, the failure of a single component may not necessarily cause an accident. Most accidents are the result of a series of component failures that have not been detected by the operator. Consequently, operators need to be constantly on the lookout for malfunctioning components which have the potential of causing lethal accidents.

There is only one method of accurately testing the low water cutoff. An operator can only accomplish this task by simulating an actual low water situation. When testing the low water cutoff, the boiler must be under pressure and firing. The water level must be lowered by gradually draining the boiler through the blowdown line.

Merely draining the float chamber of the cutoff to test the device is not sufficient. Many boiler inspectors can attest to the fact that low water cutoffs perform satisfactorily upon draining the float chamber. Still, the same cutoffs failed when tested properly.

A regular program of testing can be invaluable for safe boiler operation. Never leave the test site when performing tests. If a failure occurs, the operator must immediately discontinue the test. In the event that a low water cutoff malfunctions during testing and the water level has fallen to a critically low point, the boiler should be shut down and cooled so the water supply can be replenished. Once detected, a malfunctioning low water cutoff must be replaced immediately.

Vacuum Pump: Also called an air pump, this pump creates a vacuum in the return lines to draw condensate into the condensate tank. As the pump returns the condensate to the boiler, it vents displaced air from the tank and discharges it into the atmosphere.

The vacuum pump has a three position switch: a) a *continuous position* for testing the pump, b) a *float only position* where the pump only starts when the return tank begins to fill, and c) a *float or vacuum position* where the pump starts as in the float only mode or in the vacuum mode should the vacuum fall below a predetermined set point.

The *float or vacuum position* is the normal operating setting. Similar to the pressuretrol, the range of the vacuum switch is usually 2 to 8-inches. Translated, this means at 2-inches of vacuum, the pump starts. When it reaches 8-inches, the pump shuts off.

The vacuum created by this pump is primarily responsible for suctioning condensate out of the return lines and depositing it in the condensate tank where it becomes feedwater. The

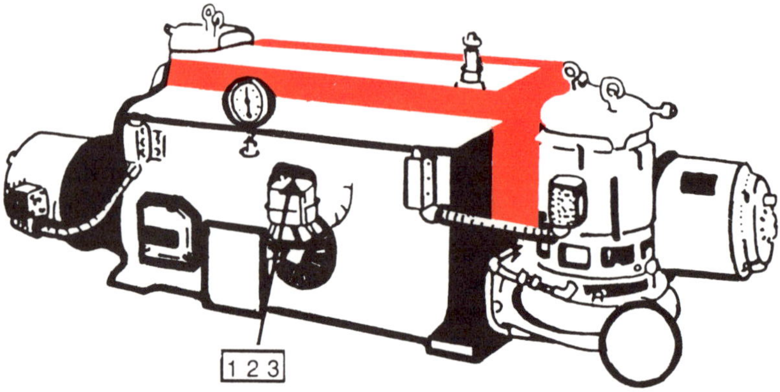

1. Continuous (a) 2. Float Only (b) 3. Float or Vacuum (c)

Figure 3-5. Vacuum pump with three position switching mechanism.

vacuum pump can only provide the boiler with the amount of condensate that is returned to the tank. If any leaks exist in the return lines, the pump cannot compensate for the condensate that is lost. This is where the city makeup water supply becomes invaluable.

Without being equipped with some switching mechanism to activate the pump, the vacuum which exists in the return lines can exert a constant draw that may eventually cause the condensate tank to flood. To avoid flooding situations, the pump switch is set in the *float or vacuum position*. At this setting, the pump can be activated either by the float or by the decreased vacuum that occurs as the tank fills.

If a boiler is taken off-line at night, the vacuum pump switch is set to the *float only position*. If any condensate returns to the tank during the down-time, it accumulates until the quantity is sufficient to raise the float and activate the pump which then feeds the condensate back into the boiler.

BOILER TRIM & INSTRUMENTATION

Boiler outlets may have threaded taps or flanged nozzles. Other openings are provided for feedwater pipe connections, drains, cleanouts and instrumentation.

Every apparatus connected to the pressure side of the boiler is **boiler trim**. The term trim usually applies to safety instruments, each of which are crucial to the operation of boilers. Briefly, trim controls the flow of steam and feedwater, relieves excessive pressure and supplies blowdown.

Consequently, it is imperative that boiler operators become familiar with every item of trim. Operators are responsible for proper maintenance and periodic testing of the instrumentation to assure safe and efficient boiler operation.

GAGES

Steam Gages: Also known as a pressure gage, this instrument indicates what the steam pressure is within a boiler. Pressure is read in increments of pounds per square inch (psi).

Every boiler must have a minimum of one steam gage installed at the highest point of the boiler's steam space. The steam gage displays the pressure of a boiler so an operator can monitor the boiler while it is on-line. Placement of the gage is critical because nothing can obstruct the operator's view of the gage glass.

The steam gage must have a siphon between the gage and the boiler. There are a number of different types, but the siphon is simply an apparatus that keeps steam from entering the gage. Steam, at extreme temperatures, can damage the delicate internal components of the gage. On initial start-up,

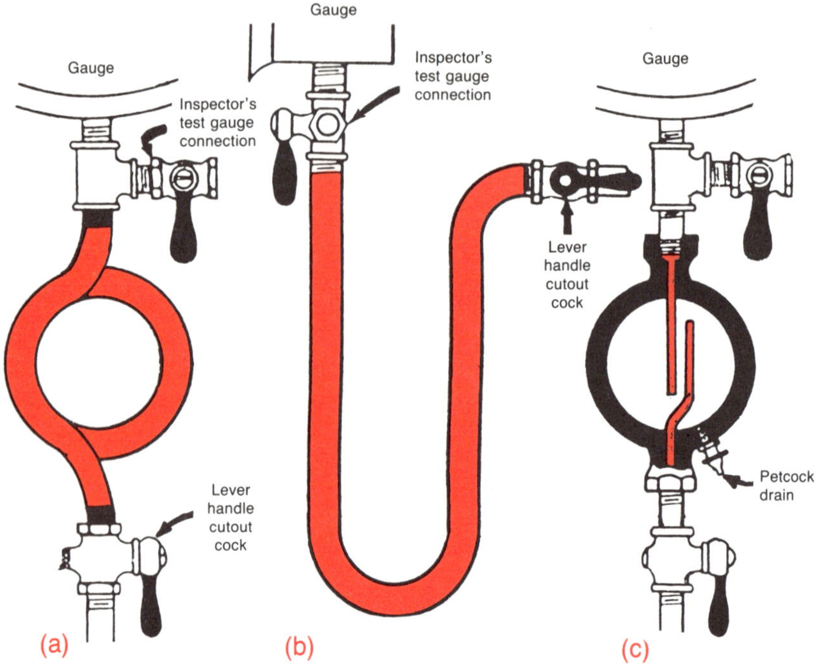

Figure 3-6. Common types of siphons installed to protect steam gages: a) pipe, b) U-tube, c) radiator.

a small quantity of steam enters the siphon and condenses into water. The water gradually accumulates. This water then serves as both a medium that transmits pressure to the gage and as a water trap that prevents additional steam from entering the gage. Without a siphon, high steam temperatures could cause erroneous readings and deterioration of the metal components inside the gage.

Steam gages are connected directly to the boiler or attached to the water column. In either case, a stop cock is installed between the gage and boiler (on the boiler-end of the siphon) to facilitate shut-off for gage maintenance or replacement. A tee fitting is also required between the gage and the gage-end of the siphon. Fitted with a inspector's gage connection, the tee permits attachment of a test gage. The test gage checks the accuracy of the steam gage. As the boiler pressure rises and falls, the readings of two gages are compared.

For temperatures that do not exceed 400°F, the steam gage dial should be graduated to 1.5 times the maximum allowable working pressure (MAWP) of the safety valve. At temperatures exceeding 400°F, the graduation should be equivalent to two times the set pressure of the safety valve.

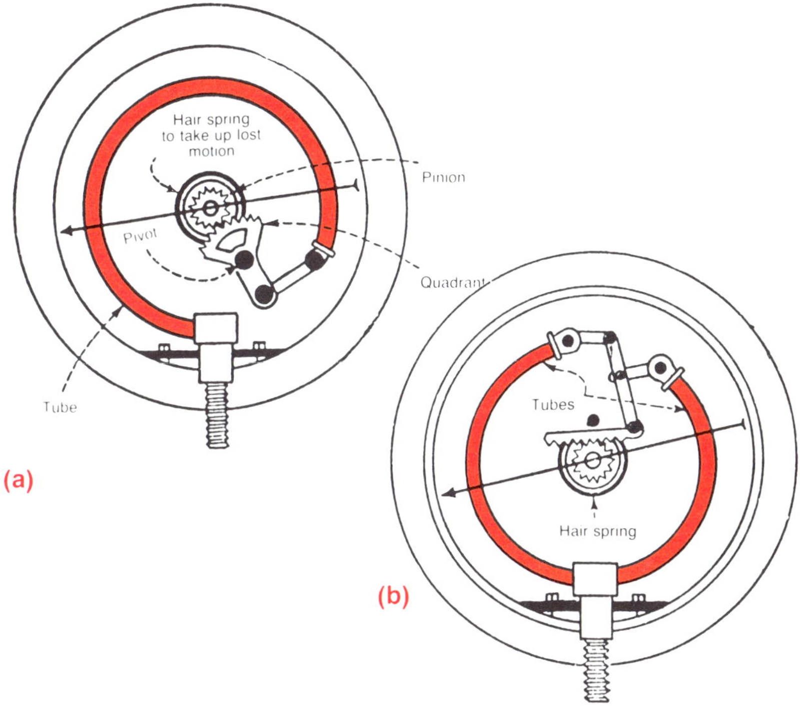

Figure 3-7. Two common configurations of bourdon tube steam pressure gages: a) single tube, b) double tube.

- Bourdon Tube Steam Gages: One prevalent type of steam gage is the single-tube bourdon gage. The bourdon tube, oval in cross section, assumes the shape of a question mark. The open end of the bourdon tube in this gage is connected, through the siphon, to the steam side of the boiler. The closed end of the tube is connected to a linked armature that ultimately ratchets along a centrally located pinion whose needle pivots over a scale calibrated in psi. When pressure rises, the tube extends lengthwise. As the lengthening occurs, the needle registers the pressure on the graduated dial.

- Diaphragm Steam Gages: Another commonly used steam gage is the diaphragm type. This gage has, as its primary component, a flexible metal diaphragm seated between a pair flanges. As boiler pressure rises, the diaphragm yields to the pressure and assumes a convex shape. The slight upward movement of the diaphragm is translated through a pinned linkage which, in turn, meshes with a pinion on the central pointer/spindle to register the pressure.

- Recording Steam Gages: This type of gage consists of a mechanism similar to either of the preceding with one major difference. A revolving chart equipped with a pen to trace a written record of pressure fluctuations replaces the stationary dial and pivoting pointer. Controlled by timers, recording gages activate automatically at preset intervals to provide comparative data over a prescribed period of time. This comparison can yield valuable information on operating efficiency and how it may relate to variables such as peak periods of usage and steam demand.

- Compound Gages: Most low pressure boilers have a compound gage that indicates system pressure on one face of the gage and vacuum on the opposite.

Water Gage Glass and Water Column: The water gage glass is an instrument that indicates water level in a boiler. In most instances, the gage glass is attached directly to the boiler shell or drum.

Both the water and steam level can be observed on a steam boiler gage glass by opening and closing a set of valves. Opening the steam valve at the high-end of the gage glass shows steam level. Opening the water valve at the low-end of the glass displays the water level.

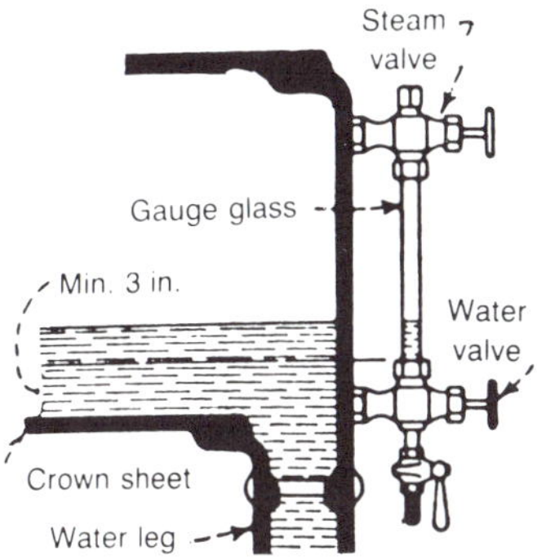

Figure 3-8. Typical water gage installation for boiler with a water-leg. Note minimum water level.

Sometimes it is necessary to use a water column to attach the gage glass when direct attachment is not possible due to design barriers.

A water column is an instrument used to divert water and facilitate attachment of the gage glass. The high-end of the water column is piped to the steam space. The low-end is piped to the water space. The column piping must meet two conditions. The steam pipe connection at the high-end of the column cannot be lower than the top of the gage glass. And the water pipe connection at the low-end cannot be higher than the bottom of the gage glass. The passage pro-

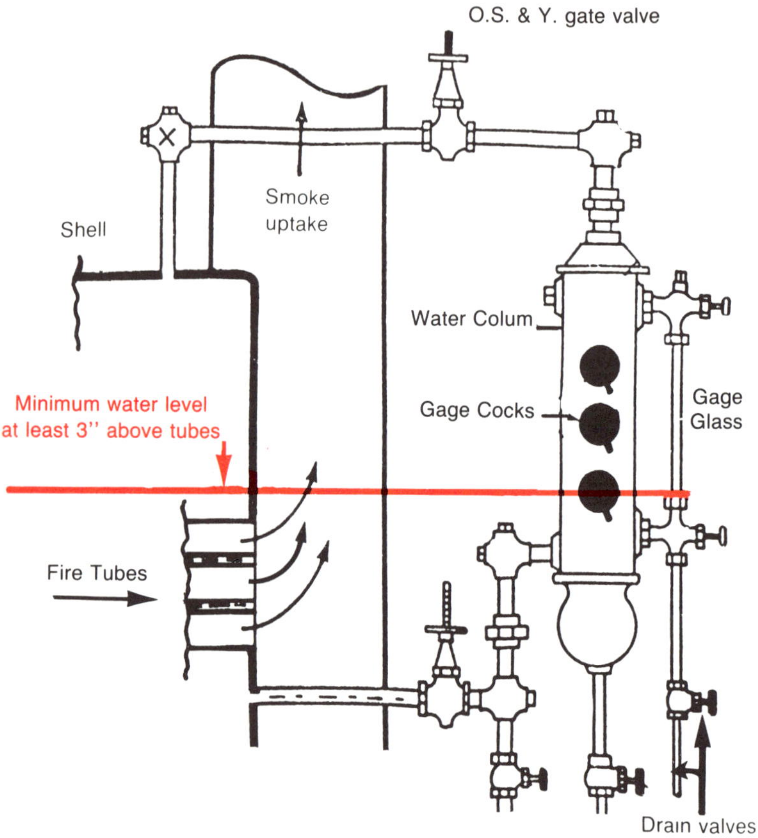

Figure 3-9. Water column/water gage combination installed on HRT boiler.

vided by the water column helps arrest water-level surges (caused by increased boiler circulation) which would otherwise show up on the gage glass. In short, the column stabilizes water-levels for accurate gage glass readings.

Water columns are simply hollow castings or forgings. At pressures under 250 psi, cast iron water columns are most common. At pressures exceeding 250 psi, the ASME Code specifies that steel forged columns are necessary.

The water column has three *try cocks* (*gage cocks*). Try cocks are actually globe valves that are back-up devices used to check water and/or steam levels should a gage glass break when a boiler is on-line. When opened, the top try cock on the column only allows steam to escape. The second or middle try cock allows both steam and water to escape. The bottom try cock (positioned level with the low-end of the gage glass) only allows water to escape.

Two styles of try cocks are common. The plain try cock has a wheel handle that opens like an ordinary valve. And the spring try cock opens by way of a pull-chain operated lever.

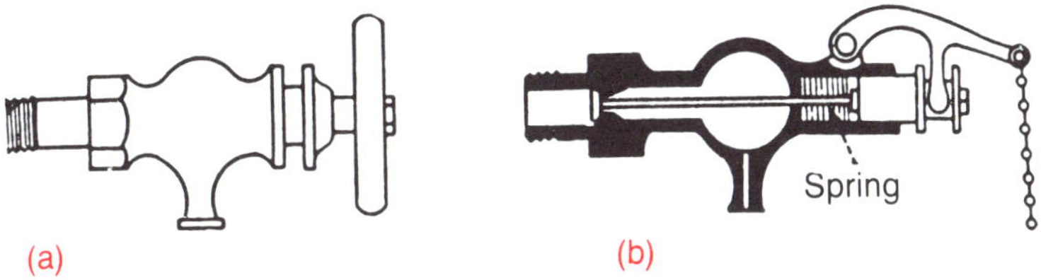

Figure 3-10. Gage cocks: a) plain type, b) spring operated.

Whether the gage glass is installed with or without a water column, the ASME code requires that the glass be mounted a minimum of 3-feet above the highest heating surface (to prevent damage by low water) and at the normal operating water level. The code also specifies the lowest part of the gage glass must show well below the normal operating water level (i.e., a minimum of 2-inches above the lowest safe water level).

The bottom of both the water column and the gage glass are fitted with blowdown (drain) valves. Gate valves are usually used for this application because an obstructed valve or pipe can easily be cleared without taking the boiler off-line for service. Opening the blowdown valves purges the passages and clears them of trapped water, sludge and sediment that may have built up.

While the boiler is on-line, an operator can use the following procedure to test whether the passages in the water column and gage glass are clear.

Step 1. Close the top valves on both the water column and gage glass. Blowdown the gage glass (open the drain valve).
- A. Water flows freely from the drain valve (go to step 2).
- B. Water flow is hampered, check and clear:
  - passage from boiler to the water column
  - passage from water column to gage glass

Step 2. Close the bottom valves on the water column and gage glass. Open the top valves.
- A. Steam blows freely from glass blowdown valve (go to step 3)
- B. No steam, check and clear:
  - passage from boiler to water column
  - passage from water column to gage glass

Step 3. Close blowdown valve on the gage glass. Open blowdown valve on the water column.
- A. Steam blows freely from the water column blowdown valve. Water column clear. **Passages clear.**
- B. No steam—proceed:
  - close column blowdown valve
  - open column and glass water valves
    - a. quick water rise to proper level (go to step 4)
    - b. sluggish water rise—obstruction remains in pipe or valves. Clear (go to step 1).

Step 4. Close both glass and column blowdown valves. Fully open water and steam valves. Seal valves on column piping in open position.

Safe Water Levels: As a boiler cools, the water level falls. While the level is declining, water can get trapped in a column if the installation is not piped properly or if passages are not clear.

Water trapped in the column can cause inaccurate readings, and ultimately, overheating can result if the water level falls too low. The following list is a general guide outlining the lowest water levels acceptable for safe boiler operation.

BOILER TYPE	LOWEST SAFE WATER LEVEL
• firetube boilers	1-inch above the top row of tubes
• SM & locomotive	1-inch above highest point of crown sheet
• vertical tube boilers	⅓ height of the tubes
• watertube boilers	6-inches above the bottom of the steam/water drum

FITTINGS

Air Cock: When a boiler is being filled with water, air gets trapped in the system. As the water level increases, the air becomes compressed. Compressing the air increases the pressure in the boiler. If measures are not taken to alleviate this excess pressure, it will invariably trip the safety valve.

An air cock is a device used to bleed air out of the system preventing pressure build-up. Essentially the air cock is a ½ to ¾-inch line (fitted with a valve) mounted on top of the highest drum of a boiler. If compressed air is not bled off through an air cock prior to firing up a boiler, the heat will expand the already compressed air and trip the safety valve(s). Air cocks are also used to bleed air out of the main header when firing-up boilers.

To purge the system of air at the main header, the air cocks are opened until steam begins to escape. Then, before opening the steam valve, close the air cock.

When warming up a boiler for operation, it is especially important to vent it with the air cock because the air can become trapped in the steam lines. Air bound lines can adversely affect the steam circulation of the entire system. In a high-rise building, for instance, the lines and radiator on each floor would have to be bled until all the air has been eliminated from the lines.

The process of purging air from a steam system is comparable to bleeding air out of the brake fluid lines on an automobile. In the case of purging steam lines, the pressure drops substantially throughout the system, as well as in the steam header and boiler itself. To compensate for the pressure drop, a boiler must generate steam well above its normal operating capacity to bring the pressure up to the correct level. Increasing the steam demand subjects a boiler to unnecessary strain and affects its efficiency. Malfunctions due to over exerting the unit in these situations are also more likely to occur.

When an operating steam generator is shut down or comes off-line for servicing, the steam it contains slowly cools and gradually condenses inside the boiler. Unless the air cocks are opened to properly vent the boiler, this cooling process creates a dangerous vacuum.

Vacuum is basically the absence of pressure. A vacuum of 14.7 psi is equivalent to a crushing or collapsing force of 14.7 pounds/in^2 of surface area. A boiler subjected to a vacuum of this extent can cause serious accidents and personal injury to operators.

Never attempt to remove a handhole or manhole cover without first venting a boiler. Removing one of these covers on an unvented boiler can cause a steam or water drum to collapse. Should a boiler collapse, the force that results is so tremendous it can physically pull the operator into the drum.

Fusible Plug: Commonly used in low to medium pressure boilers, this accessory is a device that protects a boiler from damage due to low water levels.

A small bronze or brass plug, this fitting has a tapered hollow cavity filled with a soft metal alloy, most often 100% tin. The tin filling has a melting point of approximately 450°F.

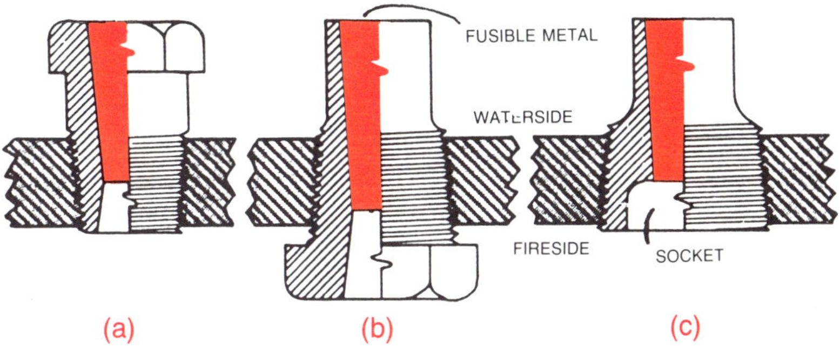

Figure 3-11. Three fusible plug designs in common use on steam boilers: a) waterside, b) bolt-head fireside, c) socket-head fireside.

Fusible plugs are screwed into the plate of a boiler at a designated level which is the highest point above the heating surface where damage can occur from low water. Normally, one end of the plug is covered by water. The opposite is exposed to heat and combustion gases. As long as the water-side of the plug is submerged and remains cool, the tin filling will not melt. If the water level drops and uncovers the plug, the filling will melt and open the passage. Once opened, steam whistles out through the cavity. The steam-kettle like warning issued by the melted fusible plug is meant to alert the operator to the low water condition before the tubes have a chance to burn.

A few variations of the fusible plug are in common use.

- Waterside plug: This plug is fastened into the plate from the waterside of the boiler.
- Fireside plug: There are two types of fireside plugs, the recessed and flush-mounted, both fastened from the fireside. They are distinguished by the way they are screwed into the plate. The flush-mounted style is run-in with a ordinary wrench. Used where a lower profile is necessary, the recessed style is run-in with an allen or plug wrench.

The waterside of any type of fusible plug must project least ¾-inch beyond the plate into the water to assure complete submersion. And the fireside of the plugs must not extend more than 1-inch from the plate surface.

Boiler pressure is always exerted on the widest end of the tapered soft metal core. For maintenance, plugs should be inspected for soot and scale build-up every time the boiler is serviced. Any build-up must be removed. If not, it can insulate the tin filling and raise its melting point. Consequently, the plug is rendered useless. It is recommended plugs be replaced at least once a year. The ASME code does not suggest used plugs be refilled.

Although fusible plugs are only practical for low to medium pressure boilers, they are widely used in coal fired and firetube boilers. The high operating temperatures and pressures of modern boilers prohibits their use because shells and drums are often too thick.

Common Fusible Plug Locations:

- locomotive type boiler—in the rear tube sheet
- HRT—in the rear head, above the upper row of tubes
- fire box boiler—in the front tube sheet
- water-leg boiler—in the highest part of the crown sheet
- Scotch Marine—in the top of the combustion chamber or in the rear tube above the highest row of tubes.

VALVES

Safety Valves: From a safety standpoint, it is mandatory to install a positive protection device on a boiler to prevent steam pressure from exceeding the pressure rating of the boiler. If excess pressure does build-up, the safety valve opens and discharges steam into the atmosphere to reduce boiler pressure and guard against an explosion.

As far as an operator is concerned, the most important fitting on a boiler is the safety valve. In most applications the only type of safety valve permissible by the ASME Code is the spring-loaded (pop) safety valve. The spring-loaded valve design is recommended because it is factory sealed and relatively tamper resistant. Adjustments for blowback (drop in pressure that closes a popped valve) should subsequently only be attempted by the manufacturer or its authorized representative.

The pop safety valve is designed to open and stay open until steam pressure drops between 2 to 4-lbs, and then close without chattering (opening and closing quickly without sticking or hesitating).

For low pressure boilers, the maximum allowable working pressure is 15-lbs psig. In other words, the boiler cannot operate safely at steam pressures exceeding 15-lbs. Once steam pressure exceeds the 15-lb limit, the safety valve "pops" open quickly to relieve pressure and prevent a boiler explosion.

Located at the highest point of the steam side of the boiler, the safety valve on a typical low pressure boiler is attached directly to the top of the boiler shell or drum. The valve must be installed in an upright position (so the spindle is vertical) to function properly. It cannot be placed in close proximity to any walkways or work platforms where personnel may be subject to scalding should the valve pop open to

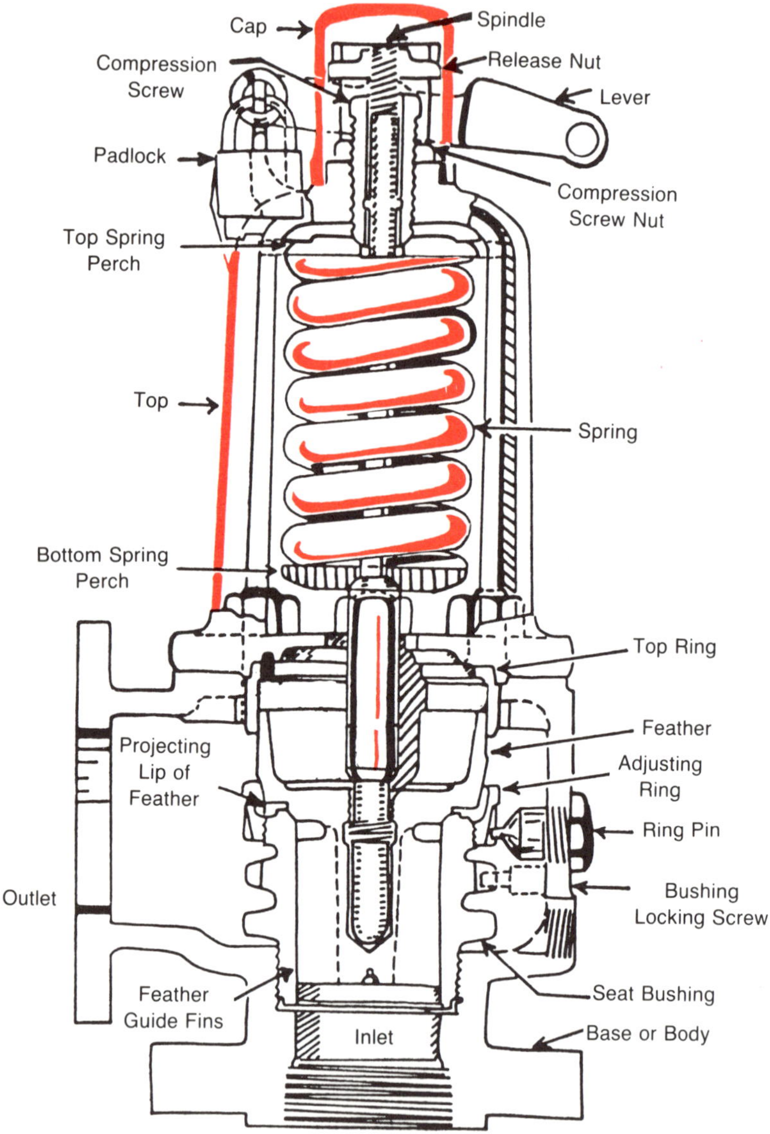

Figure 3-12. Parts of a typical safety valve.

relieve excess pressure. No valves or other fittings can be installed between the safety valve and the boiler.

Boilers are rated to operate at pressures that vary according to the size, design and capacity of the unit. Boilers rated up to 15-lbs MAWP and below are classified as low pressure boilers; boilers rated 15-lbs MAWP and above are considered high pressure boilers. The pressure rating of a boiler is significant because it determines the maximum pressure under which the boiler can safely operate. The boiler pressure rating also specifies the parameters of the safety valve. For example, if a boiler is rated for 60-lbs steam pressure, the safety valve will pop open as soon as pressure exceeds 60-lbs.

According to the ASME Code any boiler with over 500-ft^2 of heating surface must be equipped with more than one safety valve. Multiple safety valves are necessary as a precaution to guard against individual valve malfunctions.

The capacity of a safety valve must allow it to discharge the entire quantity of steam a boiler can produce to restrict pressure to an increase of no more than 6% above the maximum allowable operating pressure.

On large boilers equipped with two or more safety valves, a difference of no more than 6% can exist between the pressure ratings of individual valves. The obvious logic here is that if the first valve could not sufficiently relieve the pressure quickly enough to regulate the boiler pressure, the second safety valve pops when 6% additional pressure has built-up. If extremely high pressure build-up occurs, both valves pop.

It is possible to test safety valves by shutting off all steam outlets from the boiler and bringing the boiler up to maximum pressure to purposely pop the safety valve. An alternative test consists of simply lifting the lever manually when the boiler pressure reaches approximately 75% of its normal operating pressure.

Safety valves should always be tested manually at least once a week to avoid sticking and corrosion of the valve seat. To test a safety valve by hand without physically climbing on the boiler, ASME Code allows attachment of a block and tackle arrangement to facilitate testing. Fastened to the lever of the safety valve, a cable passes over pulleys so that the weight of the cable will not exert any pull directly on the lever. An operator can then merely pull on the cable at ground level to open and test the safety valve.

Testing a safety valve can also be accomplished by bringing the boiler up to pressure, shutting off all steam outlets from the boiler and forcing the boiler to exceed its maximum allowable pressure rating. The safety valve(s) should then pop to prevent the pressure from rising further. In no case should the operator allow the pressure to rise more than 6% above the MAWP of the boiler.

An operator can calculate the steam pressure or total force acting against the safety valve, as follows:

Given: 3-inch valve base and a MAWP of 15 psi
Area of valve = $D^2 \times$ Constant (Fixed value)
where D^2 = 3-inch valve base (3×3)
and Constant = 0.7854
Area of valve = $(3 \times 3) \times 0.7854 = 7.07$ in^2

Total force required 7.07×15 psi = 106-lbs
to raise the safety valve off its seat.

Blowdown Valves: Additional water must be added to boilers periodically to maintain the proper water level. As water is added, the concentration of sediment and soluble salts gradually increases unless the water is treated. Both soluble and insoluble material can accumulate within a boiler and hamper operating efficiency if not blown out of the system on a regular basis.

The process of purging a boiler of foreign matter is called blowdown. Every boiler must be equipped with valves to allow blowdown.

Typically, a screw type gate valve is used for a blowdown connection. More often than not, as a safeguard, a quick-closing lever operated gate valve is installed between the boiler and screw gate valve to provide a positive seal. This intermediate valve is positioned to absorb the majority of the wear incurred by the connection and should always be opened first and closed last during blowdowns.

There are rigid ASME code restrictions for valves used on blowdown connections. Globe valves cannot be used because their design allows sediment build-up. Gate valves function best where valves are subject to little operation. Cast iron valve bodies are permissible under pressures of 100 psi. Over 100 psi, steel valve construction is required. Two valves must be used if pressure exceeds 100 psi, and one of the two is required to be of the slow-opening variety. All blowdown valves must be rated at a minimum of 25% greater pressure than the MAWP of the boiler.

Location distinguishes the two basic categories of blowdown connections, surface blowdown and bottom blowdown. Although not found on all boilers, the surface blowdown valve skims off impurities floating on the surface of the water inside the boiler. Located at the normal operating water level, the surface blowdown permits removal of impurities that cause foaming. Foaming is caused by high surface tension due to scum building up on the surface of the water. This condition prevents steam bubbles from breaking through the water surface and consequently hinders steam production.

The bottom blowdown serves a number of functions:

a) High water control. If the water in a boiler rises above the normal operating level, it begins to take up a portion of the steam space. Increasing the quantity of water in a boiler significantly slows the heating capability of a boiler and proportionately increases its fuel consumption. In addition, as the water level rises, carry-over and priming may occur. This condition is attributed to incomplete separation of steam and water. When the steam area of a boiler has been reduced by a rising water level, the steam generated does not have sufficient room to separate entirely from the water to produce clean, dry steam. Rather, the steam hovering over the water surface tends to wick up water and impurities then carries both over into the steam lines. Water saturated steam traveling in the steam lines at high velocity causes water hammer which can eventually weaken joints and loosen connections. When saturated steam tries to pass through a sharp piping turn such as a 90-degree, for example, it slams into internal surfaces and causes lines to vibrate. As a result, the line can burst or rupture from continual hammering and vibration. In extreme cases, the steam lines can become waterbound and/or clogged by impurity deposits which inhibits circulation.

b) Sludge and sediment removal. To rid the boiler of corrosive waterborne impurities held in suspension and to keep them from settling on tube surfaces, water needs to be chemically treated. The chemicals help separate impurities, such as scale forming salts, from the water. The impurities precipitate as heavier than water solids (sludge and sediment) and fall to the bottom of the boiler. Before blowdown, the boiler's heating process must be suspended temporarily to halt water turbulence and allow the solids to settle out. Once settled, they collect in the mud drum. The solids (mud) can then be purged from the boiler through the bottom blowdown valve.

The blowdown valve is located in the lowest point of the waterside on a watertube boiler and in the rear bottom of a firetube boiler. On the watertube boiler there may be one or two blowdown valves to blow off the mud drum(s). The valves are located on either the front or back of the boiler.

Discharge from the blowdown cannot be routed directly into the sewer due to the high water temperature. Rather the discharge is piped through a line to a tank or pump where it is allowed to cool before going into the sewer.

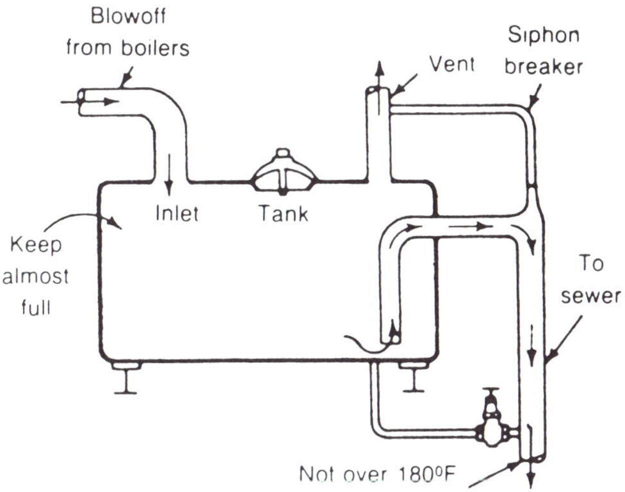

Figure 3-13. Blowoff tank system used to cool boiler water prior to discharging.

c) Chemical concentration control. Untreated water contains impurities that can damage a boiler. Certain minerals and salts which form scale or corrode metals occur naturally in water. For instance, scale insulates heating surfaces such as boiler tubes. The insulation inhibits water from sufficiently cooling the tubes. If the heating surfaces are not cooled quickly enough, the tubes can overheat, blister and eventually burn out.

As steam is generated, the impurities held in suspension by the water become concentrated because the original volume of water has been reduced. Even over a short duration, the continual process of introducing fresh, untreated makeup water (along with its impurities) into the boiler ultimately raises the concentrations. The higher the concentration of these impurities, the more corrosive it becomes to the internal boiler components.

The corrosive nature of concentrated acidic impurities wears away tube surfaces, plates, seals, gaskets, rivets and stays (supports) weakening the entire internal structure of the boiler. Boiler water is treated at regular intervals with chemicals to prevent such corrosion. Water impurities such as scale forming salts, minerals, and acids are chemically converted into a non-adhering sludge that remains suspended in water so it can be purged at low fire through the bottom blowdown valve.

d) Dumping. To entirely empty or dump a boiler, it has to be taken off-line. Firing is completely discontinued and draft is cut down so the boiler cannot cool too rapidly. When 1 or 2-lbs of pressure remain on the steam gage, the air cock(s) are opened to prevent vacuum formation. After the boiler has stopped steaming, the main steam stop-valve is closed. The boiler is then ready for dumping.

The boiler should be cleaned thoroughly on both the water and the fireside at least once a year. It is also convenient to inspect the boiler after it has been dumped and cleaned and prior to putting it back on-line.

STEAM REGULATION APPARATUSES

Steam is capable of efficiently transporting large quantities of heat from a boiler to a heat exchange unit with only a minimal loss of heat. It flows naturally from the high pressure of the boiler to the lower pressure in the steam mains and circulates faster through piping than any other fluid system. Steam promotes its own circulation due to the lower pressure along the steam mains and in the heat exchangers caused by pipe friction and the condensing process as it loses heat. This natural tendency eliminates the need for a mechanical process to maintain circulation.

Pipe systems need to be properly designed and installed to avoid unnecessary pockets where condensate can accumulate in amounts sufficient enough to block the circulation of steam. Pockets are formed by improper pitch, sags caused by lack of support to ensure a continuous pitch or improperly installed fittings.

Steam piping systems are not liable to leak as easily as are fluid systems. When steam leaks do occur, damage is minimal because cubic foot of steam only condenses into a small quantity of water. Much of the escaping steam often evaporates into the atmosphere before it even has a chance to condensate.

It is essential for every boiler operator to comprehend the fundamentals of steam circulation and how flow control fittings affect the performance of any steam generator.

VALVES

To control the flow of steam from the boiler to the header, every boiler must have a main steam stop-valve. The restrictive valve designs are not well suited for this application. So, due to their design, gate valves are used almost exclusively for main and auxiliary steam outlets where the valves must open and close completely to form positive seals. The advantage of the gate valve lies in mechanics of its wedge

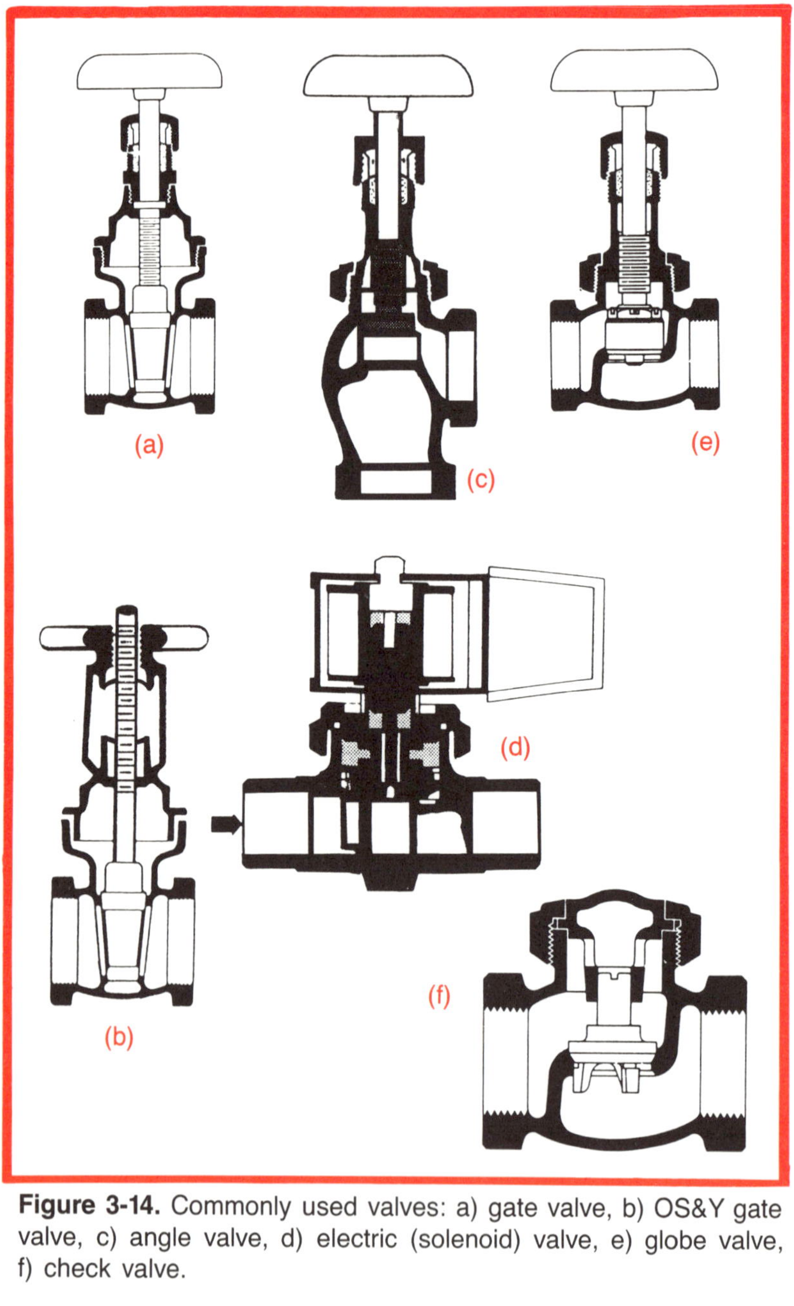

Figure 3-14. Commonly used valves: a) gate valve, b) OS&Y gate valve, c) angle valve, d) electric (solenoid) valve, e) globe valve, f) check valve.

principle operation. In the fully open position, steam flow is totally unobstructed. Since gate valves have a straight-through flow, no damming effects occur to trap particles of sediment. The gate valve also offers the advantage of opening slowly. Its stem needs to be turned a number of times to raise or lower the plate across the passage in the valve body. This slow operation prevents hammering and its harmful affects on the lines.

There are two types of gate valves, the non-rising stem and inside screw, and the rising-stem and outside-screw and yoke (OS&Y). The OS&Y type is used most commonly. The position of the rising stem of the OS&Y makes it easily discernible whether the valve is open or closed. If the major portion of the threaded stem protrudes and is visible above the handwheel, the valve is open. If little of the stem is seen, it is closed. This valve is very useful in hard to reach areas. An operator can merely observe the position of the stem to determine if it is open without having to actually test the valve by opening and closing it by hand.

One drawback to the plate style gate valve is that it cannot be left partially open. The throttling that results from the steam being obstructed by a partially closed plate causes chatter and vibration which wears excessively on the seating surfaces of the valve. Once the surfaces are worn, positive shut-off cannot be assured. Consequently, the damaged valve must then be replaced.

When a boiler reaches operating temperature, after going back on-line, its main steam gate valve needs to be opened slowly to gradually blend the steam already in the header with the newly generated steam. An operator should always open the steam stop-valve, or close the steam stop valve, to the fully opened of fully closed position to avoid the damaging effects of throttling.

For cases where a single boiler is harnessed in series with other boilers an automatic non-return valve is often employed on mains in conjunction with an OS&Y stop valve. A non-return valve is usually installed on outlets to aid a boiler in building up steam pressure. Non-return valves can be closed by manually screwing down the outside stem but can only be opened by boiler steam pressure. For example, if a boiler is about ready to cut-in or go on-line, the OS&Y stop valve is opened. As soon as the boiler pressure rises slightly higher than the pressure in the steam main, the boiler pressure raises the non-return valve shut-off and steam automatically enters the header. Consequently, if the boiler pressure drops below the header pressure, the non-return valve closes and cuts-off the steam in the boiler. The primary function of the non-return valve is to promote steam flow from the boiler to the main but to inhibit (check) the reverse.

The valves used for most common steam or water piping applications are the gate and globe valves. Globe valves get their name from globular shaped valve bodies. Globe valves are used where flow control (throttling) of steam or water free from suspended particles or where tightness to prevent leakage is necessary. The flow through a globe valve moves in a Z pattern that allows throttling control without excessive wearing of the seat and disc. Chattering is eliminated because the Z pattern forces the disk of the valve against the flow instead of across it such as in a gate valve. Even so, throttling does take its toll eventually, but globe valves are serviceable and are ideal where frequent valve maintenance might be required.

The fact that the critical components of the globe valve (seats, discs and packings) are easily accessible and lend themselves to replacement gives the globe valve an advantage over gate valves. To replace worn components, pipeline pressure is eliminated, the valve bonnet is unscrewed and the parts replaced. The actual body of the valve does not have to be disconnected from the rest of the piping for servicing. Still, globe valves have restricted passage. Where unobstructed passage is essential, gate valves need to be installed.

STEAM TRAPS

Steam traps are automatic valves designed to increase boiler efficiency, without the loss of steam. The function of a steam trap is to drain-off condensate from the steam distribution system. Anytime steam moves from the boiler to the header and enters the system, its temperature drops slightly and produces condensate. Condensate not only slows circulation and reduces efficiency, it also creates water hammer as previously mentioned in the description of carryover.

In review, water hammer occurs whenever condensate mixes and travels with steam moving at high velocity. Whenever the water hits a critical bend in the line, an audible bumping or hammering results. To eliminate hammering and vibration, steam traps are placed wherever condensate buildup might exist: at the ends of the header, off each branch line and at radiator, convector or heat exchanger exits—anywhere steam gives up its heat.

Essentially, steam traps can be categorized into three groups.

- Thermostatic traps. This group of steam traps operates in response to a thermal element that is sensitive to the difference in temperature between steam and condensate.
- Mechanical traps. This group of traps is buoyancy responsive. In other words, mechanical traps respond to the raising or lowering of a float or a bucket controlled by the condensate level.

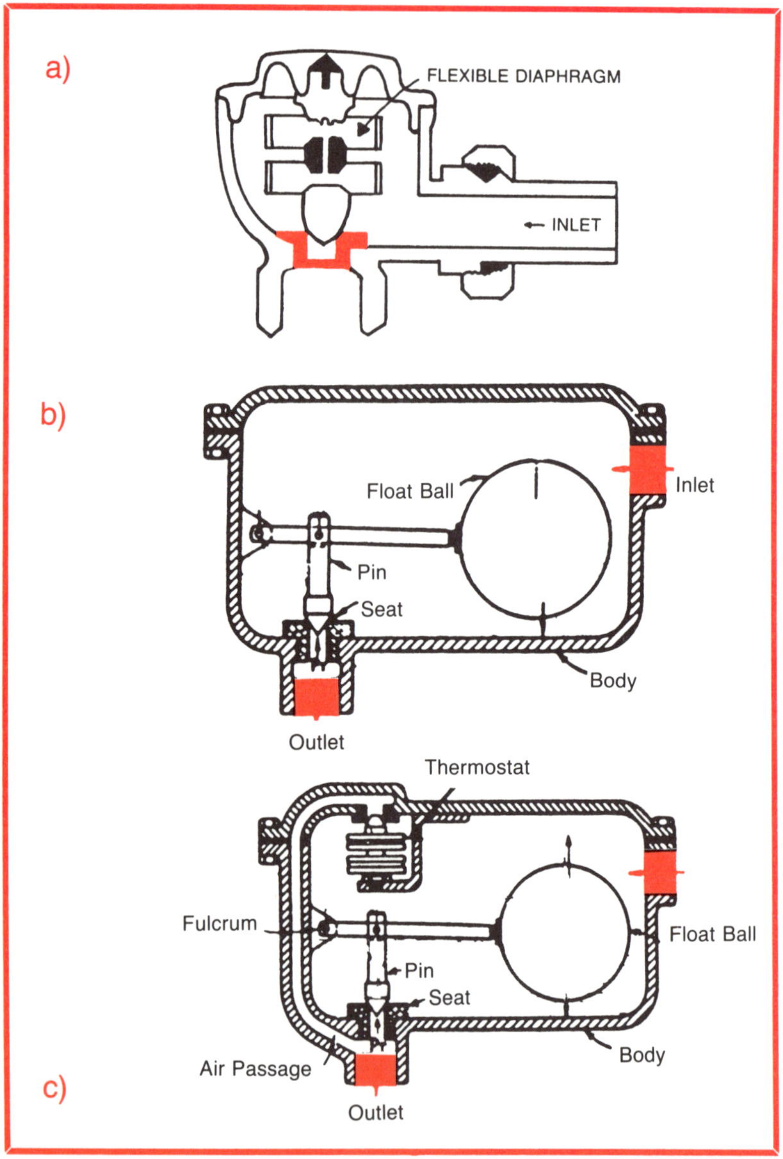

Figure 3-15. Steam traps: a) thermostatic, b) conventional float, c) float and thermostatic trap.

- Thermodynamic traps. This group responds to change of phase attributed to flash steam that is produced by the condensate as it flows from a higher to a lower pressure.

Non-return traps consist of any trap design that has no live direct steam connection from the boiler. These simply discharge condensate as waste or divert it to a hot water receiver against little or no back pressure.

Some of the widely used non-return traps are:

Inverted Bucket Trap. The most common, this mechanical steam trap is designed for a wide array of sizes, capacities and operating pressures. Its popularity can be credited to both its flexibility and simple internal construction. In this trap design, a bucket float responds to the level of the condensate entering the body of the trap. The buoyancy of the bucket acts on the lever mechanism to open and close the discharge port.

Prior to operation, all inverted-bucket traps must be primed or filled with water. When primed, the bucket has no buoyancy.

During operation, the inverted bucket is entirely submerged in condensate. As steam displaces the water from inside the bucket, the bucket becomes buoyant (floats) and closes the discharge port. While the condensate continues to flow into the trap, the steam in the bucket slowly condenses. Air and non-condensable gases escape through the bucket vent and accumulate above the bucket. Once air and non-condensable gases are vented, condensate can flow into the body of the trap. The gases are finally expelled from the trap when the bucket loses buoyancy, sinks and opens the discharge port.

Condensate is then discharged until steam fills the inside of the inverted bucket and displaces the condensate making the bucket buoyant again. It rises and closes the discharge port to complete the cycle. The trap operation is always intermittent or cycles either open or closed.

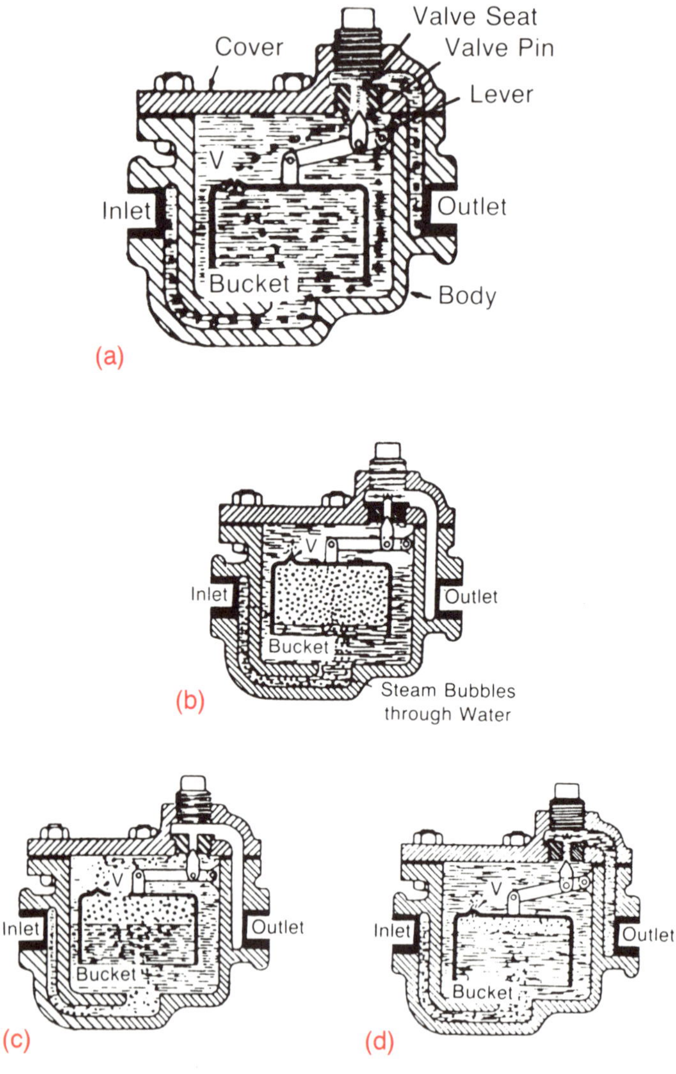

Figure 3-16. Operation of inverted bucket trap: a) steam first turned on, b) condensate flow becomes normal, c) trap operating, d) end of cycle.

In many instances, bucket traps can even vent air and non-condensable gas accumulations without interfering with condensate discharge.

The capacities of inverted bucket traps are large in comparison to traps that operate in a throttled position to allow continuous flow. A large capacity is advantageous because it can be adjusted to suit the actual equipment performance after the trap has already been installed. The flexibility afforded here makes the availability of a variety of valve-seat-port diameters that cover a wide range of operating pressures. When necessary, required modifications can conveniently be made in the field.

An added advantage is that temperature has no affect on the operation of this style trap. Also, the inverted bucket trap withstands shock loading, is self-purging and efficiently removes condensate as soon as it is formed.

Inverted bucket traps are less prone to malfunctions due to rust, scale and other abrasive materials that occur in the piping than are other designs. Still, any mechanical trap should be protected by a dirt strainer installed directly ahead of the trap inlet.

A few disadvantages also occur with the inverted bucket trap. Over-sizing may cause loss of prime, and the inverted bucket is not recommended for super heating applications. The inverted bucket is not self draining, and it requires a change of the valve and seal for different operating pressures. This type of trap is most often used with autoclaves, cooking kettles, fuel oil preheaters, unit heaters and steam mains.

Thermostatic Traps: Commonly used in conjunction with convectors or radiators, thermostatic traps are widely used in space heating systems. The use of these traps is not confined to heating applications because they are designed for a variety of commercial and industrial process applications.

Basically, this type of trap operates on the temperature differential that exists between steam and condensate. Thermostatic traps are designed to function according to temperature drops confined within a limited range between 5°-30°F.

- Balanced-pressure Traps. A number of thermostatic trap designs exist, but the balanced-pressure is by far the most common. The body of the trap consists of a thermal element which is either a bellows or a flexible diaphragm containing a mixture of water and alcohol. The element is constructed to allow expansion and contraction and is vacuum sealed.

    The port seat is open at start up because the liquid in the bellows or diaphragm element is contracted. When heated condensate or steam enters the trap, the liquid in the element vaporizes and expands the bellows or diaphragm closing the discharge port in the valve seat.

    When the discharge port opens, only condensate (not steam) leaves the radiator or convector. In the event of failure the trap will fail in the closed position, preventing any loss of steam. If the trap does fail, condensate can flood the equipment until the temperature inside the trap drops and permits the discharge to open. To avoid flooding, cooling legs are employed. This device is simply an adequate length of pipe installed between the return outlet and the trap inlet. The cooling leg stores and sufficiently cools condensate allowing the trap to discharge before flooding occurs.

    The capacity of balance-pressure element traps depends on the trap travel (lift of the valve) and the diameter of the discharge port. The valve lift is governed by the expansion and contraction of the element that is dependent upon temperature drop. The shape of the valve itself determines the time factor involved with reaching maximum capacity. A flat seating valve reaches its maximum capacity position with less travel than V-shaped or spherical valves. There is an advantage in using the latter shapes over flat-seating valves. They open slower and offer a positive seal to prevent discharge port leakage.

Additional advantages of the balance-pressure element trap is that its single valve and seat operate over an entire range of pressures. It is also self draining, when piped with a bottom outlet, vents substantial volumes of air and operates over wide condensate load fluctuations. Oversizing is not economically sound, but a trap that is too large will still function efficiently and offer good service.

These traps are subject to shock loading damage and often require a cooling leg to prevent condensate flooding. They are not recommended for superheating steam applications. Dirt poses trouble for this trap design, and it becomes more acute at low pressures accompanied by a light condensate load.

- Float Thermostatic Traps. This trap design consists of a float that opens and closes the discharge valve as a result of the amount of condensate that accumulates in the bowl of the trap. Condensate causes the float to rise and open the discharge valve. Suction and pressure remove the condensate, and the float drops and closes the valve. Air in the trap is expelled by a thermostatic valve at the top.

Return traps are designed to return condensation to the boiler, against boiler pressure, rather than discharge it as waste. *Tilting traps* and some forms of *floating traps* are commonly used for this purpose. They are usually installed a few feet above the boiler and steam is run from the boiler to the trap. Tilting of the bowl or movement of the float opens the discharge valve simultaneously opening another valve to let steam enter the trap. Pressure from the steam line, together with the few feet of gravity head, causes condensate to run from the trap down to the boiler.

Traps are generally the most neglected components in the heating system. If for example, the trap on a radiator or convector becomes waterbound, the heating process stops.

If the trap sticks open, steam can blow through the radiator or convector and flash the condensate in the return lines back into steam. Once converted back into steam, the condensate cannot be pumped back to the boiler. Eventually, the vacuum pump would become steambound. As a result, overheating can occur due to the low water level in the boiler.

There are a few ways to test a trap to determine whether it is functioning properly. One is to install a pair of strap-on thermometers. Place one on the steam line ahead of the trap and one behind the trap. A 10° or 20°F difference should be apparent between the two readings. At the first location steam is entering the trap; and at the second location steam is leaving the trap. If steam is blowing through the trap, the temperatures would be the same on both sides. Another method utilizes a temperature stick which is just a special crayon that has a controlled melting point. Mark the discharge side of the trap with the crayon. If it melts, the trap is stuck open and steam is blowing through. This condition indicates the trap is not working properly.

In order to size and select the proper trap, contact a manufacturer's representative who can provide the information an operator requires.

SECTION 3

SYSTEM COMPONENTS

Review Questions

1. What apparatus prevents backflow on feedwater lines? How does it operate?
2. What device is used to overcome boiler pressure so feed-water can enter and replenish the system water?
3. For which piping applications are globe valves best suited and why?
4. The low water cut-off safeguards against overheating conditions. List the common types of cut-offs and briefly describe each.
5. What device is used to maintain a constant water level and how is it activated?
6. Automatic devices are failsafe, true or false?
7. What is the safest way to test the low water cut-off?
8. In what mode must the boiler be in to test the low water water cut-off?
9. If a low water cut-off fails during testing, what procedure must an operator follow?
10. How is condensate drawn back into the condensate tank?
11. The cycling of a vacuum pump is controlled by a switch, list the three settings of the switch.
12. Describe how flooding is controlled in the condensate tank
13. What instrument monitors steam pressure and where is it located?
14. Identify the apparatus that keeps steam from entering the steam gage. Describe how it functions.

15. How is the accuracy of a steam gage tested?
16. How should the steam gage be graduated in relation to temperature and the safety valve?
17. List the two most prevalent types of steam gages and briefly describe how each operates.
18. Describe the function of the gage glass.
19. At what position must the gage glass be mounted to prevent damage by low water?
20. What apparatus is often employed to facilitate attachment of the gage glass and what are its two primary functions?
21. Try cocks are back-up devices that check water/steam if a gage glass breaks. Identify the two most common try cocks and detail how each operates.
22. List the lowest safe water level in the four boiler categories
23. What fitting functions as a bleed-off for air entrapped in a boiler? Why is it important and where is it located?
24. Identify and describe the fitting which safeguards a boiler from low water levels. How does it work?
25. What device protects a boiler from excessive pressure, where is it typically located?
26. Define MAWP and the pressure ratings that distinguish low pressure boilers from high pressure boilers.
27. What is chattering and what does it indicate?
28. Under what condition are multiple safety valves required? At what percentage over the MAWP should additional safety valves be rated?
29. Briefly describe how a safety valve operates.
30. Explain the two methods of testing safety valves.
31. How often should safety valves be tested, and what con ditions warn of a possible malfunction?
32. Detail how to determine the force required to "pop" the safety valve.

33. What apparatus is used to drain, purge or maintain proper boiler system water level? What type of connection is typically employed for this application and what is its minimum rating?
34. Identify the type of valve installed between the boiler and and its drainage connection. Why is this valve design used for this purpose?
35. What distinguishes the two types of blowdown connections Name the two types and enumerate the functions of each.
36. Why can't globe valves be used for blowdown connections? Where are they best employed?
37. Where does sludge and sediment collect for blowdown?
38. What valve design is used almost exclusively for main steam stop connections? Why?
39. Why is an OS&Y type used most often on steam lines?
40. What valve is commonly used in conjunction with the OS&Y valve? Why?
41. What drawback is associated with plate style gate valves?
42. Briefly describe the advantages of using globe valves on steam and water piping applications.
43. What apparatus is designed to increase boiler efficiency without losing steam? State its function.
44. Explain water hammer.
45. Enumerate the types of steam traps.
46. How do return and non-return traps differ?
47. Identify the most common non-return, mechanical trap employed in steam systems. List the attributes which lend it such wide use.
48. Where are thermostatic traps primarily used? List and briefly describe the two variations of the thermostatic traps.
49. Name the traps that return condensate against boiler pressure.
50. What happens if a trap malfunctions?
51. How is a trap tested? What is used to conduct the test?

4

FUELS & COMBUSTION

SECTION 4

FUELS & COMBUSTION

COMBUSTION PRINCIPLES

The primary purpose of a boiler is to transfer the heat of combustion to the water or steam to be heated. The boiler operator must then have a fundamental understanding of the phenomena called combustion and how it takes place in the boiler.

The most rudimentary starting point is the flame. When a combustible gas combines with atmospheric oxygen and is heated to its ignition or kindling temperature, a chemical recombination of basic elements occurs. The reaction produces heat and light in the form of a flame. A nonluminous flame is hotter than a luminous (clear) flame due to the fact that a greater portion of the chemical energy emitted from the reaction is transformed into heat energy.

Basically a high speed oxidation process, combustion is the rapid burning of fuel and oxygen which, in turn, liberates heat. The heat produced by the reaction continually relights the unburned portion of the fuel and sustains a constant burn.

There is a significant difference between the burning properties of various fuels. The type of fuel used in a furnace also determines the quantity of air required for combustion. Air in excess of the required quantity is necessary for complete combustion to occur. The volume of excess air required depends on factors such as fuel properties, size and design of the combustion chamber, maximum furnace temperature and thoroughness of the air/volatile gas mixing.

An inherent problem with supplying excess air to assure complete combustion is that operating efficiency is reduced to a certain degree. The excess air lowers the furnace temperature and absorbs heat that should more properly be utilized for generating steam.

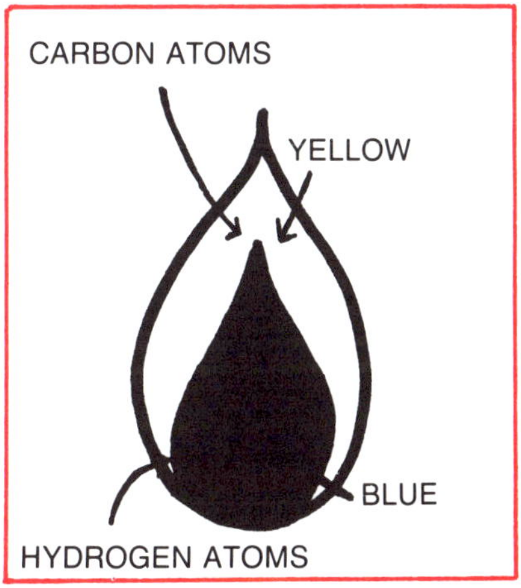

Figure 4-1. In a typical flame, hydrogen burns first and then the carbon particles burn.

Approximately half of the combustion air (primary air) is introduced into the furnace along with the fuel. The remaining portion (secondary air) is introduced over the fire or at the burner. The primary air controls the combustion rate, and the secondary air assures complete combustion.

Furnace temperature influences the fuel ignition and the combustion rate as well. In order for combustion to take place, the mixture of air and fuel combustibles must be raised to its ignition temperature. Once the firing point of ignition is reached, the boiler furnace must be capable of rapidly supplying oxygen in the required amount to sustain complete combustion. When furnace temperature exceeds what is required for ignition, temperature is relatively insignificant because it exerts little influence on the oxidation process of combustion.

To briefly summarize:

- **Complete combustion** is the burning process where a quantity of fuel is totally consumed in an oxygen rich environment.

- **Incomplete combustion** occurs when the entire fuel supply is not consumed during combustion. Incomplete combustion creates carbon monoxide and unburned carbon particles which are often evidenced by grey smoke and soot build-up. Insufficient mixing of air and volatile gases, lack of combustion air or cooling of combustibles can all hamper complete combustion.

- **Perfect combustion** cannot be achieved under normal boiler operating conditions. Perfect combustion can only be attained in a laboratory situation where air/fuel mixtures and ignition temperatures can be controlled with high precision.

Boiler operators are responsible for maintaining complete combustion in on-line boiler furnaces to maximize operating efficiency and minimize air pollution. Always adjust the air/fuel mixture so a greater air to fuel ratio is used. A hotter fire will result, fuel consumption will be reduced and harmful by-products from incomplete combustion will not be discharged into the atmosphere.

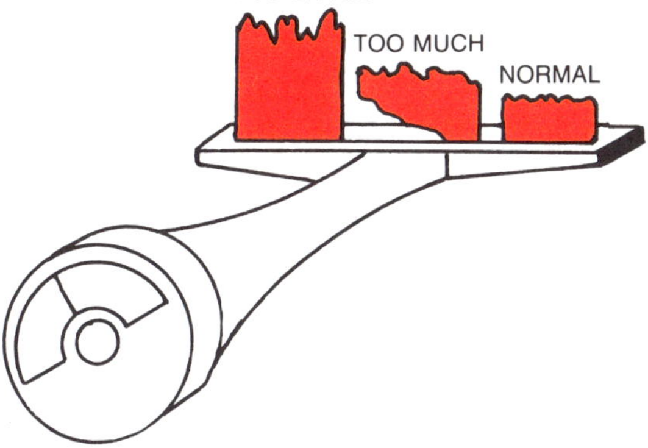

Figure 4-2. Flames resulting from improperly adjusted primary air.

COMBUSTION CONTROLS

As technology has become increasingly more sophisticated, many advances have been realized in automated boiler controls. The use of automatic controls has gradually allowed manual operation of various boiler systems to be phased out. The advent of packaged control components has relieved operators from certain tasks, such as ignition and burner-flame control, that are easily handled with more precision by automated controls. At the same time, operators can redirect their attention to tasks that can decidedly increase the overall efficiency of the boiler.

Designed to automatically regulate the fuel feed in relation to the correct ratio of air required for complete combustion, combustion controls respond to steam-pressure changes.

Primary Pressuretrols: Three basic types of combustion control systems are commonly employed: *on/off controls*, *positioning response controls* and *metering type controls*. When steam is not needed they will shut down the system. If unsafe conditions occur they lock the boiler out of the power unit and will not allow the boiler to fire off at all.

The **on/off control** is the simplest, least expensive type of control but its use is confined to firetube and small watertube boilers. When the steam pressure drops to a preset level, a mercury switch or pressurestat is activated to automatically feed an air/fuel mixture into the furnace, ignite the burner and open the main air damper. The control shuts off when the boiler attains a preset pressure limit.

When the on/off control is in the "off" mode, natural draft can move through the furnace and absorb heat as it passes through. This heat loss can lower the boiler's efficiency.

Commonly employed on packaged boilers, the **positioning control** tracks steam pressure and automatically adjusts or directs fuel and air flows to a predetermined alignment, usually varying the primary and secondary air by damper adjustment. This control facilitates shifting from high to low fire to meet fluctuating steam demands. Still, the positioning control may not detect variables such as soot build-up or power surges at the blower motor that can affect airflow. Even though the positioning control can maintain a constant steam pressure, it may not be capable of sustaining high combustion efficiency due to manual adjustments necessary to compensate for load changes.

When a boiler operates under fluctuating loads for extended periods, a **metering control** lends the best precision for controlling combustion. Capable of maintaining the optimum air/fuel ratio over a wide spectrum of load ranges, the

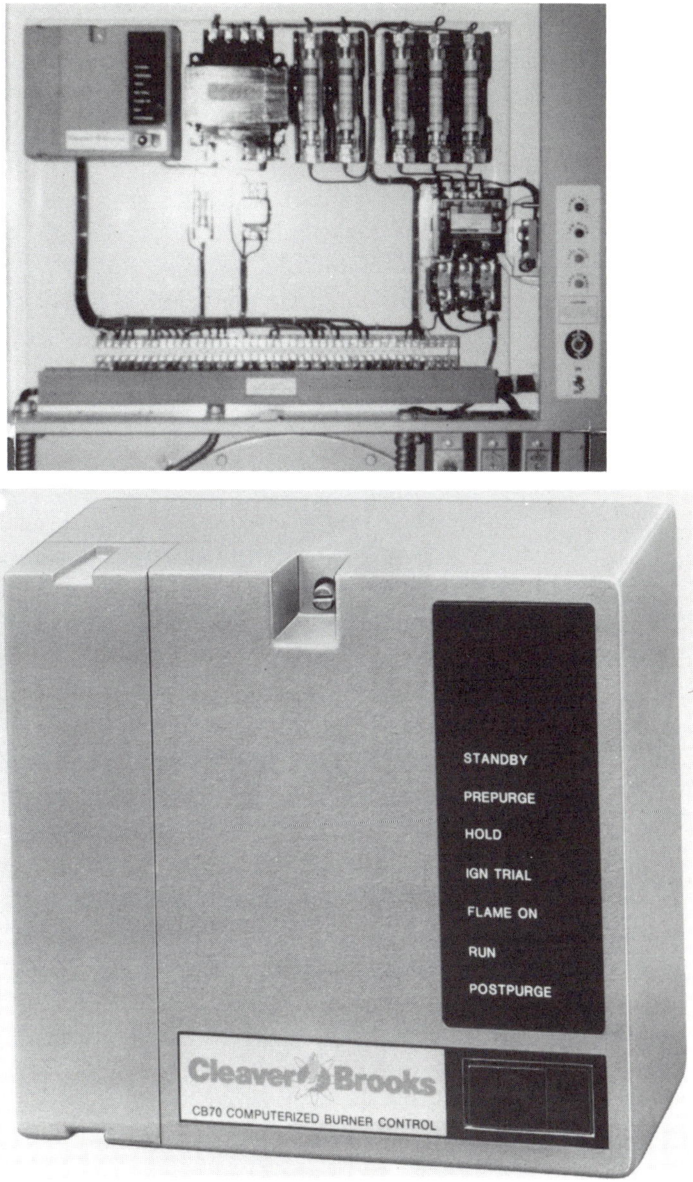

Figure 4-3. Computerized burner control. *(Courtesy, Cleaver-Brooks)*

metering control actually measures fuel and air flow, as well as the draft loss. This feature allows the metering control to independently modify valve and damper positions to maintain measured flows instead of responding to preset adjustments as in the positioning control. This feature proves valuable in helping compensate for variables such as fuel quality, degree of boiler fouling and wear in mechanical devices that may impair operation.

A variety of combustion controls with either hydraulic, electric or pneumatic relays that transmit signals to predetermined control points are now in common use.

Programming Clock: The most important control on boilers, the programming clock literally safeguards the boiler through its operating cycle. When a boiler first starts, the programmer purges the boiler (eliminates any unburned fuel or gases in the fire box before and after lighting of a boiler). By means of a rotating cam or a computerized mechanism, the programmer then lights the pilot, either with an electric spark or an igniter, which in turn lights the main burner.

When steam pressure drops to the pressuretrol setting, the modulating pressuretrol completes an electrical circuit to the programming clock that starts the cycle. After completing the program, the pressuretrol and modulating pressuretrol start the burner from low fire to high fire. When the pressure reaches its high set point, the burner goes to its lowest set point and shuts off. The programming clock then goes into its purge cycle and shuts off to await a restart.

The sensory mechanism of the programmer (the fire-eye) is a safety device that prevents furnace explosions by controlling pilot and burner ignition. The burner control system must have a method of detecting whether or not ignition has been accomplished and a flame established and maintained during the on cycle. If a flame has not been established, it is necessary to stop the fuel supply and turn off the burner. This flame detection task is left to the fire-eye.

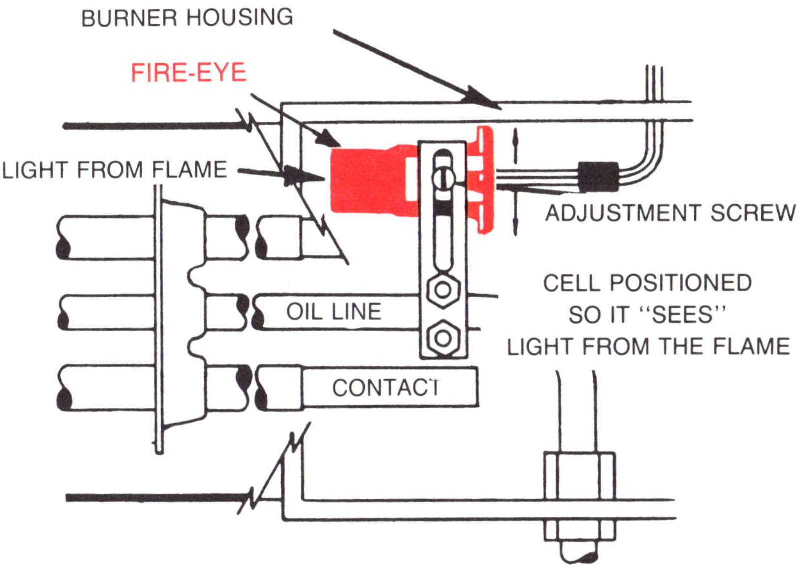

Figure 4-4. Proper positioning of fire-eye sensor.

Positioned so it can see the pilot and main burner at all times, the fire-eye requires precise alignment at exactly the right angle to the pilot and main burner so it can actuate protective controls on the fuel feed in case the pilot light is extinguished.

The main monitoring component of a typical fire-eye is a lead sulfide cell that is approximately the size of a small glass marble. The sulfide cell has the unique capability of conducting electricity when it detects light in the visible range of the spectrum. When the cell detects light, it conducts enough electricity to transmit signals to a sensitive relay in the primary control. The transmission indicates that ignition has occurred and fuel can continue to flow to the burner. In the dark, the cell will not conduct electricity. Consequently, the primary control will not allow fuel to flow to the burner.

The fire-eye supervises the flame throughout its entire burning cycle. If the fire goes out for any reason, the electrical transmission from the fire-eye stops and the fuel valve is automatically shut off by the primary control. A purge cycle is then initiated to clear the furnace or fire box of all combustible gases before pilot ignition is attempted and the cycle restarted.

Placement of the cell is critical. The cell must see light from the flame but it cannot mistake transient light from other sources such as an open inspection door for a flame. On installation, the cell is carefully positioned by the manufacturer and its location should not be disturbed.

Alternate types of flame detectors are sometimes used, but the majority of boilers in operation today utilize a fire-eye sensory scanner.

FUEL & FUEL SYSTEMS

Fuel can be burned in one of two ways, either in suspension or on a fuel bed. A fuel's ability to burn is known as its flammability. Actually, flammability is the inherent ability of fuel to be converted into a gaseous state to make it combustible.

Burning gas is an obvious example of suspension burning. In order to burn both coal and oil in suspension, both fuels must be reduced to small particulate matter in order to expose maximum surface area and to prepare the fuel so it can be mixed with an adequate amount of primary air. With coal, the reduction process is pulverizing; with oil, atomizing.

A boiler operator must understand that coal and fuel oil are combustible but they are heavy and must vaporize into gaseous hydrocarbons in order to burn. For instance, prior to burning, oil must be reduced to small droplets which then need to be atomized into even finer sized particles so the flames of a burner can *crack* the particles and reduced them to their most basic molecular elements—hydrogen and carbon. Oil that has not been reduced or atomized, can be ignited with a flame but it would quickly burn itself out due to a lack of oxygen. The same principle applies to the use of coal.

Once the fuel particle/air mixture is ignited, the action of the resulting flame must produce enough turbulence to thoroughly mix oxygen (derived from secondary air) with the burning fuel particles and gases to sustain complete combustion. Turbulence not only helps mix fuel and air, it also washes away the spent layer of combustion product gases surrounding the fuel particles revealing a fresh strata of unburned fuel to enhance combustion. Without turbulence, burning particles become shrouded in a layer of involatile combustion by-products that will not burn and support further combustion.

If an adequate amount of oxygen is lacking in the air/fuel mixture, it is considered rich. A fuel rich mixture produces a long, lazy flame that smokes, evidence of incomplete combustion. The flue gases given off by fuel rich flames contain unburned carbon particles that accumulate as soot and carbon monoxide. If an excess amount of oxygen is present, the mixture is considered lean. Lean mixtures produce shorter, hotter flames and an overall cleaner fire due to complete combustion.

The fuels used to power boilers can be classified into three main categories: solids (wood and coal), liquids (fuel oils derived from petroleum) and gases (natural gas, producer gas and blast furnace gas). The fuels which burn readily and liberate the most heat are those with a high hydrocarbon content.

Many boilers burn gas exclusively, and most use natural gas. Approximately 99% methane, natural gas is an excellent fuel. Compared to manufactured gases, which are by-products of other substances and have high hydrogen concentrations, natural gas is a relatively slow burning gas. The more hydrogen a gas contains, the faster it burns. The faster gas burns, the higher the consumption rate.

The wide use of natural gas to fuel boilers can be accounted for due to the fact that it is composed almost entirely of gaseous hydrocarbons. Since natural gas already exists as an elementary fuel in particle form and is ready to burn, it does not require further break-down for burning as is the case with coal or oil. Not only is natural gas convenient to use, it burns cleanly and is economical. A burner equipped to use gas requires no additional devices to atomize or pulverize fuel in preparation for burning.

GAS FUEL SYSTEMS

In its natural state, gas exists as an elementary fuel (i.e., no further reduction is required to prepare it for combustion). It is also a fuel that offers convenience in that it is readily available and requires no special storage facilities. Consequently, natural gas has gained wide popularity as a primary fuel source for boilers, especially packaged boilers.

The pressures in city natural gas lines may vary substantially from region to region. To compensate, boiler gas mains are routinely equipped with a governor or balanced zero reducer to lower the city gas line pressure to zero—before the gas enters the boiler. Commonly installed just ahead of the governor, is a vaporstat—a pressure operated switch that is

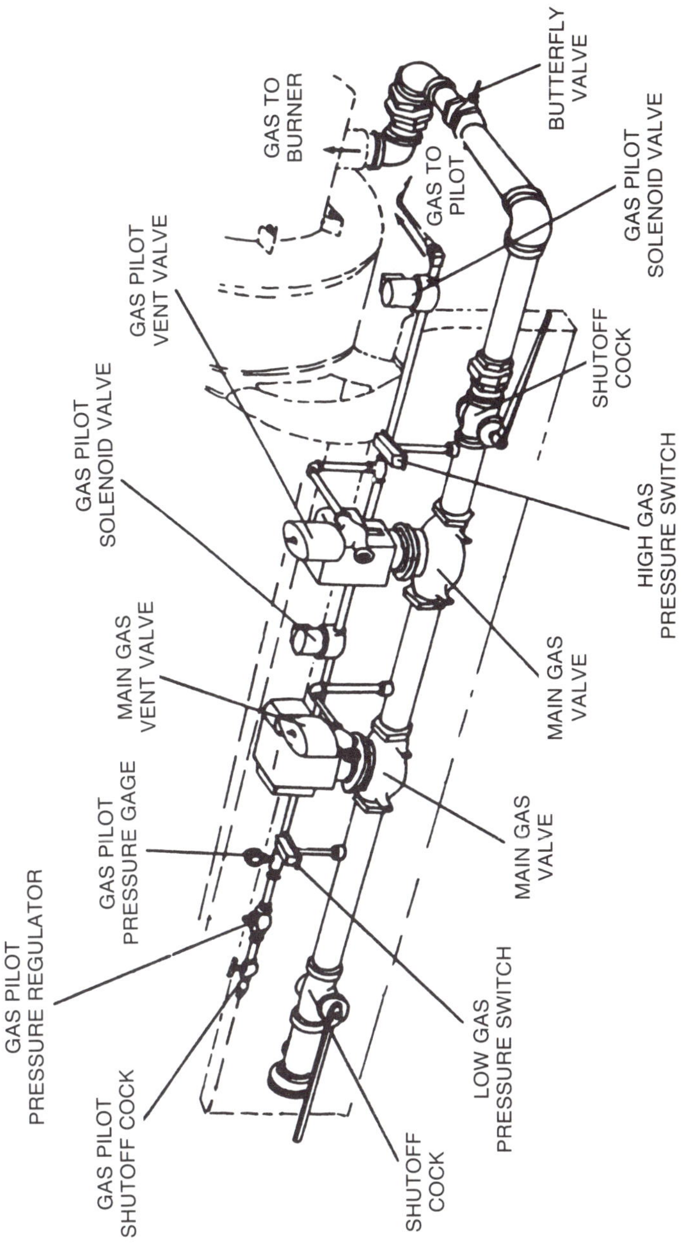

Figure 4-5. Typical gas train. *(Courtesy, Cleaver-Brooks)*

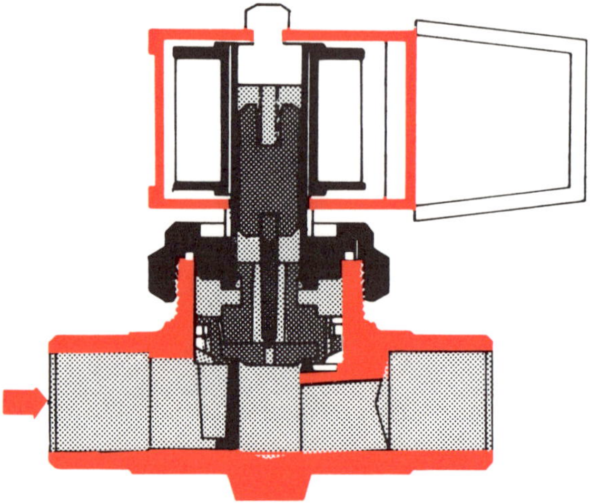

Figure 4-6. Electric valves or solenoids are installed as safety shut-offs.

actuated by the presence of gas line pressure. In the presence of pressure, the vaporstat resumes the *on* position and allows gas to fuel the main burner.

The gas main of a typical burner is also fitted with a solenoid switch (electric valve) that diverts gas to fuel the pilot burner and a manually operated stop-cock used to shut off gas flow to the burner if repairs are necessary. As a safety precaution, the head of gas main is also equipped with an electric valve designed to close the line if the pilot does not light properly and remain lit. Once the pilot lights and remains lit, the valve automatically opens to fuel the main burner.

In the common ring burner, a central manifold routes gas to a number of small single jets or circular rows of jets. Briefly, gas is discharged from the jets through an inner ring of ports drilled into the hollow core of a block of heat resistant material. The gas mixes with a stream of air that passes through the hollow core of the block. Many ring burner designs are are fitted with louver type doors that promote air turbulence and also regulate the air flow through the core of the burner block.

The amount of gas flowing into the burner is determined by the size of the gas orifice and the manifold pressure. The orifice is a small nozzle with a hole that controls gas flow. The larger the hole, the faster the gas is allowed to flow. Attached to the manifold, the nozzle fits in a small hole in the center of the burner. The size of the orifice is determined by the specific gravity of the gas, the heat value of the gas and how much gas each orifice is required to supply. Elevation of the installation also affects the orifice size. Regardless of the gas used, at elevations above 2,000-feet, gas flow is reduced by approximately 4% for every 1,000-feet above sea level.

Low pressure burners, using gas at a pressure less than 2 psi, are usually multi-jet configurations. Multi-jet designs help to thoroughly mix air/fuel.

High pressure burners require specially modified air and gas mixers to compensate for the increased gas line pressures. The high velocity of the gas is used to draw air into a mixing chamber (venturi tube) with a constricted center that controls the entrainment of air and fuel. The fuel nozzle and air inlets can be adjusted so the mixture delivered to the gas burner tip is correct and requires no supplemental air.

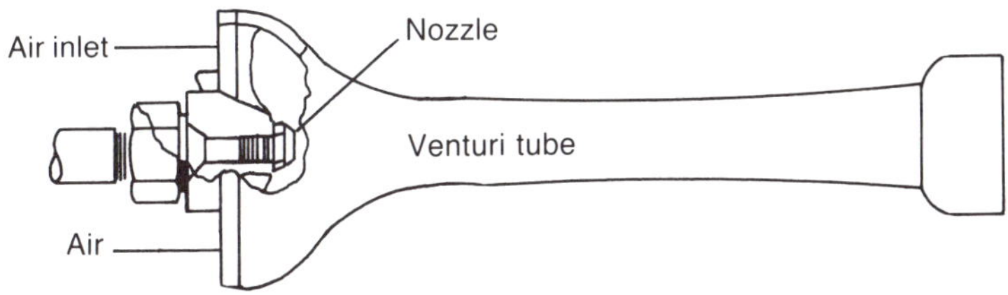

Figure 4-7. Venturi tube is required on high pressure burners.

Correct air/fuel mixture at the tip of a main gas burner is necessary to sustain complete combustion. Designed to provide constant, even flames, main burners (with the exception of high pressure burners) are required to supply adequate primary and secondary air to assure maximum combustion efficiency. The main burners must also facilitate the regulation of primary air. As previously detailed, primary air is brought into the burner and mixed with fuel before ignition occurs. Secondary air is supplied from outside the burner and mixes with burning fuel at or near the burner ports.

When the main gas solenoid valve opens, gas instantly enters the burner, passes through a regulating valve and enters a mixing chamber. Suction caused by the fuel flow pulls primary air into a mixing chamber (mixjector). An increase in the primary air, which becomes entrained with the fuel, improves burner performance because the resulting combustion is less dependent on secondary air.

In well designed burners, the secondary air reaches the burner ports at an even rate providing a good air/fuel mixture. As the burner heats up, the secondary air required to complete combustion enters the burner by means of natural draft. On larger installations, where an increased amount of secondary air is required to complete combustion, it is often necessary to employ a forced draft. In this case, a draft blower forces air through a motorized control valve to the burner cage. (Most cages are equipped with an adjustable ring to control secondary air.) As the air/fuel mixture passes through the cage, it is ignited by a smaller burner, the pilot.

Connected to the main gas valve via a piece of small diameter tubing, the pilot burner requires only a small quantity of gas to operate. The pilot burner is normally mounted so the tip of the pilot flame is directed towards the main burner for ignition. In this position, the flame also surrounds

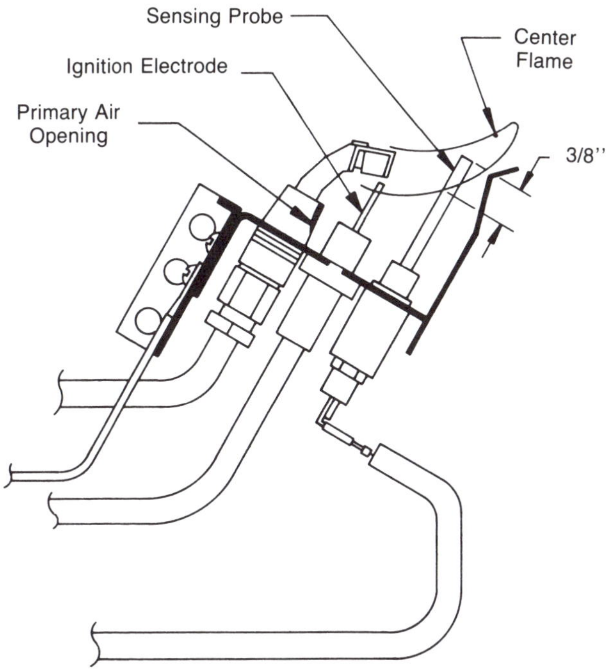

Figure 4-8. Typical electrically ignited pilot burner with thermocouple safety. Note: if flame is yellow, primary air slot may be blocked by dirt or lint. *(Courtesy, Burnham Co.)*

the tip of the pilot safety (thermocouple or flame rod). This safety is usually an electric device that determines whether or not the pilot is lit. The heat generated by the pilot flame at the tip of the thermocouple sets up an electric current that passes through an electromagnet. The field of the electromagnet opens the pilot gas valve and allows gas to flow. If the pilot goes out, the electric current is broken and the pilot gas valve closes the gas main to safeguard against explosions.

In a cold furnace, flames can be smothered if a burner has insufficient primary air. In this instance, the flame lifts off the burner and puffs out, often extinguishing the pilot. On

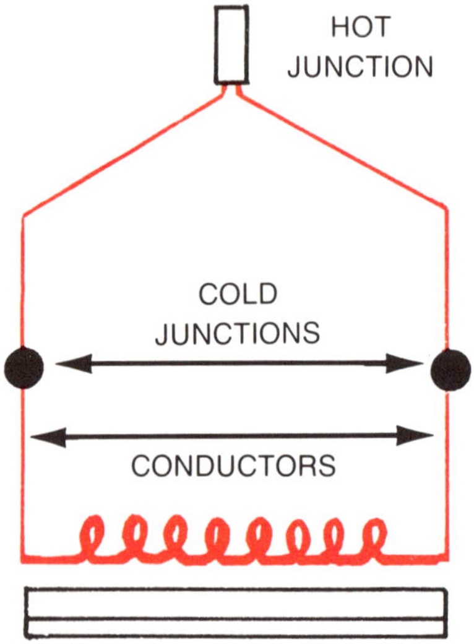

Figure 4-9. Simple schematic showing thermocouple operation.

the other hand, an excess amount of primary air produces a loud popping when the burner shuts off.

When a burner shuts off, a slight draft still exists. The draft continues to draw air into the burner where it mixes with the remaining fuel. The mixture ignites immediately. Under most circumstances, the resulting *mini* explosion has adequate force to extinguish the pilot.

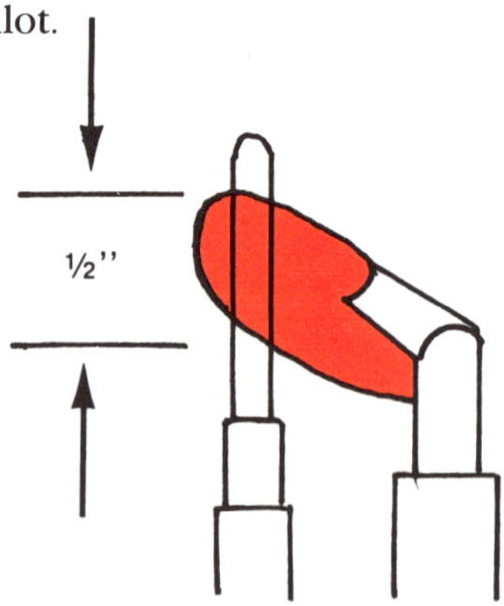

Figure 4-10. Pilot flame should properly surround approximately ⅜ to ½-inch of thermocouple depending on model and manufacturer's specifications.

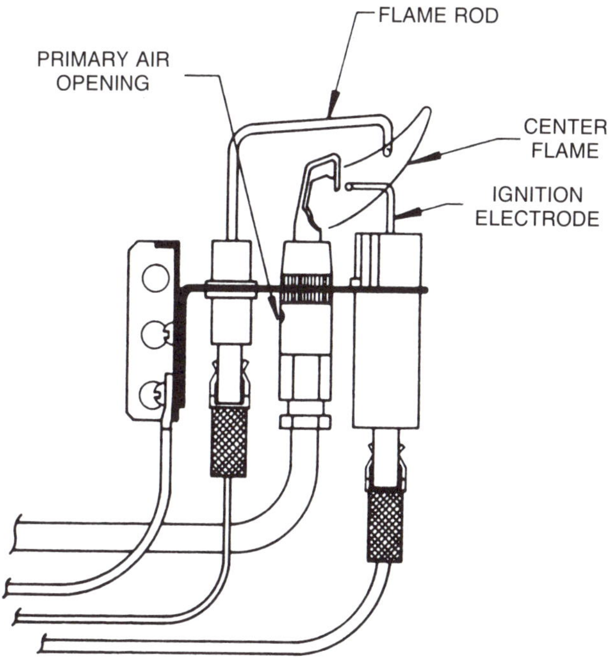

Figure 4-11. A normal flame on a pilot burner equipped with a flame rod. *(Courtesy, Burnham Co.)*

Lighting sequence: Close the main switch that controls power to all electric controls on the boiler. Open the pilot gas cock and push the pilot ignition button on the control board. The pilot will light. Next, manually open the main gas cock and reset the safety shutoff valve by lifting the handle. The gas then flows up to the balanced zero reducing governor and assumes zero pressure as it passes through the governor. The gas pressure switch (vaporstat) completes an electric circuit as soon as it *proves* gas pressure to the governor.

The vaporstat *proves* gas pressure by means of a diaphragm and switch. As pressure builds up in the gas line, it acts against the diaphragm in the vaporstat. As the diaphragm expands it tilts a mercury switch which completes the electric circuit to the forced draft fan. The completed circuit then starts the forced draft fan. Allow the fan to run for a short interval before lighting up so any gas accumulations in the furnace can be cleared.

Installed along the air line, the motorized air control valve (a slow moving electric motor) operates a damper type valve to control air flow. In the low fire position, it allows minimal air flow. Only a small quantity of air under pressure passes through to the mixing chamber. As the air/fuel mixture passes through the venturi and into the cage of the main burner, the pilot ignites the mixture at the block.

After the main burner has been lit and the boiler warms up, turn the the selector switch to automatic and push the run button on the panel board to place the burner in the high-fire mode. Once accomplished, the motorized air control valve gradually opens the damper valve to increase both the air flow and its velocity. As the air flow and the speed with which it flows increases, the flow of gas into the mixjector increases proportionately. With the boiler in high-fire, the burner becomes subject to control by the modulating pressuretrol and the on/off pressuretrol.

Note, the pilot must ignite before the manually reset safety shut-off valve will open. The thermocouple (flame rod) and fire-eye must prove the pilot flame and complete the electrical circuit to allow the safety shut-off valve to open. In the event of a pilot or main burner flame failure, the valve will close and shut off the gas flow. The safety valve is also wired in series with the low water cut-off so that it will close in the event of a low water condition.

OIL FUEL SYSTEM

Oil fired boilers use several different grades of oil which are classified according to their viscosities. Viscosity is used to describe the physical property of an oil which is its resistance to flow (i.e., its thickness). The thicker an oil, the more resistant it is to flow and the higher its viscosity. For instance,molasses has a higher viscosity than ordinary vegetable oil.

Oils are graded according to their viscosities: #2 oil is a thin oil like diesel, #4 oil is a little thicker, #5 oil is very thick and #6 oil, when cold, resembles tar. The higher the viscosity, the more Btus/gallon the oil contains. For example, #2 oil has a Btu content of 144,000 whereas #6 oil contains 152,000 Btus.

When using fuel oils, an operator must be able to distinguish between two important temperatures: the **flash point** and the **fire point**. The flash point is the temperature at which the volatile elements in the fuel will flash when exposed to an open flame. The fire point is the temperature at which the oil will burn continuously.

Remember bulk quantities of oil must necessarily be stored in tanks, and those are usually located outdoors. As the outside temperatures drop, the viscosity of oil changes. It thickens and flows less freely. In very cold weather, outside storage tanks could deliver cold oil to the nozzle of the burner. Cold oil results in poor atomization. In fact, the oil may be so cold that it cannot be ignited at all. And even if it can be ignited, it produces long, narrow, noisy flames that burn on the back wall of the combustion chamber, rather than directly in front of the burner, causing impingement.

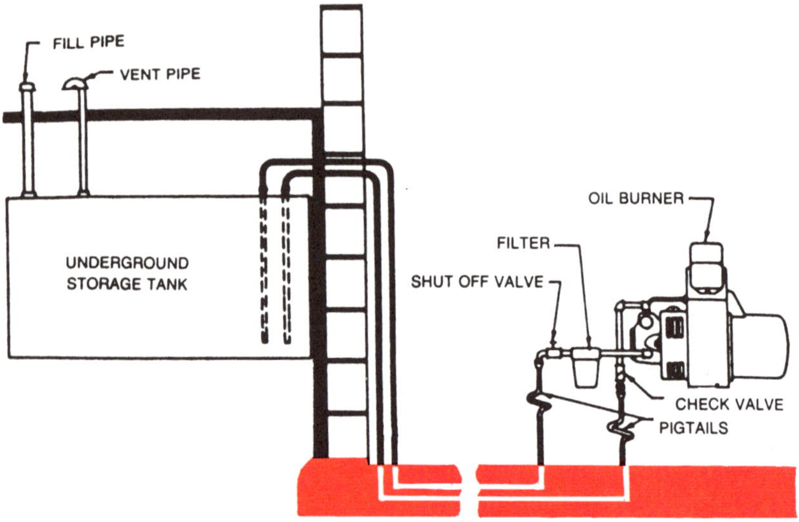

Figure 4-12. A two pipe system is commonly used where vertical lift exceeds eight feet.

To prevent the effects of pumping cold oil to the burner, tanks which store heavier oils such as #5 and #6 need to be heated. Increasing the temperature of the oil stored in the tank reduces the viscosity of the oil and allows it to be pumped to the burner in a preheated condition.

Oil flows from the storage tank to the pump through a suction line equipped with a suction gage, an oil temperature thermometer and a duplex strainer to filter impurities out of the oil.

The pump has a relief valve that returns excess oil to the supply oil line if upstream pressure becomes excessive as indicated by the pressure gage. From the pump, the oil flows to a steam heater and, if necessary, to a supplementary electric heater.

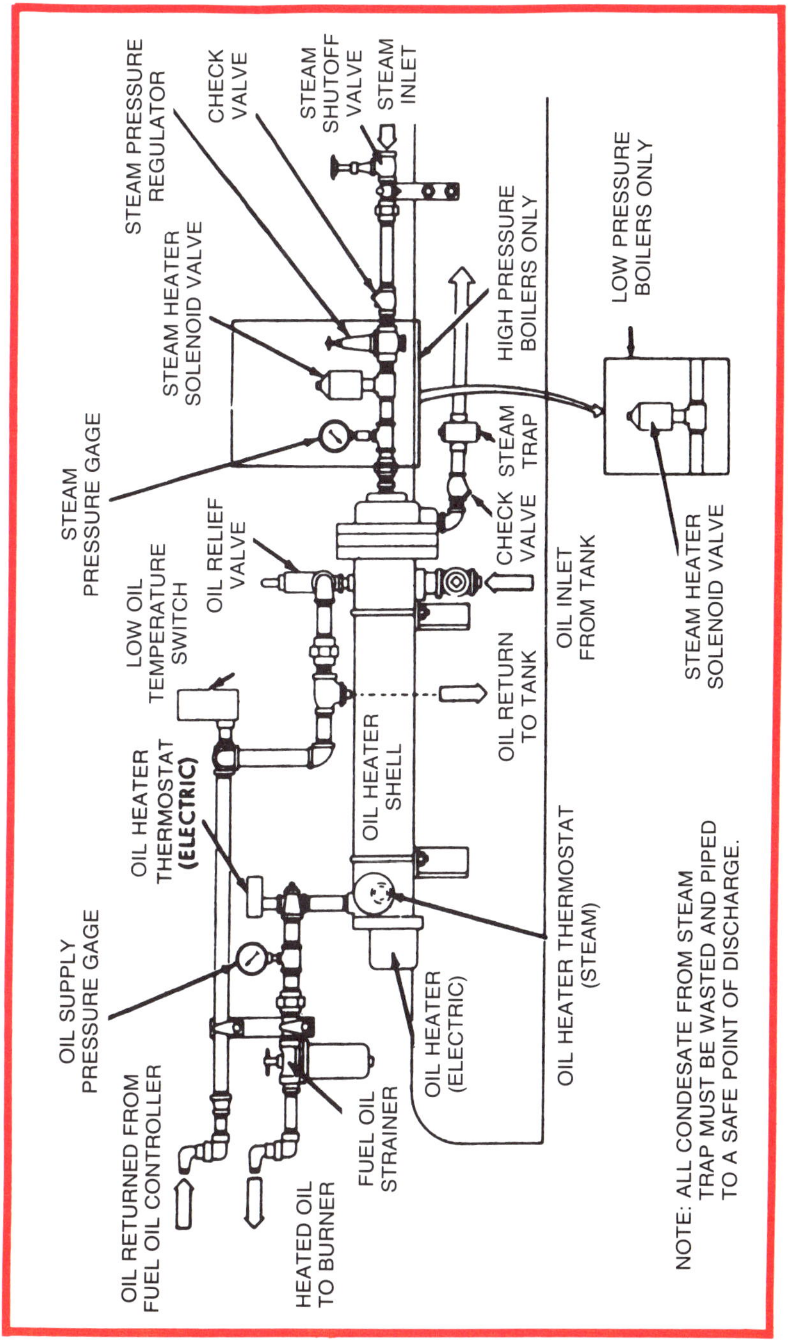

Figure 4-13. Typical steam oil heating assembly. *(Courtesy, Cleaver-Brooks)*

The line connecting the electric heater to the oil burner is equipped with another duplex strainer, an oil pressure gage and an oil pressure regulator.

Usually, the oil burner is supplied with 10% more oil than required. The excess is returned to the storage tank for recycling.

Air in the oil line can cause the flame at the burner to pulsate and/or go out altogether. To assure that no air enters or becomes trapped in the oil line during repairs or maintenance, components such as the pump, steam heater and electric heater are isolated by stop valves to facilitate taking them off-line.

FUEL OIL SYSTEM COMPONENTS

Fuel Oil Preheaters: Because #6 oil (Bunker C oil), in particular, is so thick, the tank in which it is stored must be heated to between 100-120°F by a steam coil or an electric heater before the oil can even be pumped to the burner. From the tank, the oil is pumped to a steam heater where the oil temperature is raised nearer to its flash point. The oil then moves through another electric heater that raises the oil temperature to its fire point (150-180°F in a rotary cup burner, 180-200°F in an atomizing burner).

The preheaters used with #6 oil are usually shell-and-tube steam heaters and electric heaters. Usually located near the burner, the electric heater is used when insufficient steam is available for preheating, particularly at start-up.

The shell-and-tube steam heater has a bulb that senses temperature fluctuations. When temperature increases, gas contained within the bulb expands and places pressure on a diaphragm. This pressure overcomes a countering spring pressure and throttles the flow through the steam valve. When the temperature cools, the pressure on the diaphragm drops and the spring pressure opens the valve which allows more steam to enter the heater. Stop valves are located on each side of the control valve to isolate the unit for maintenance and repairs.

Duplex Strainers: Most modern oil systems are equipped with duplex strainers mounted on the suction line ahead of the pump. An increase in the suction gage reading indicates that the strainer is either fouled or the oil entering the strainer is cold and clogging has occurred.

Figure 4-14. Duplex strainer.

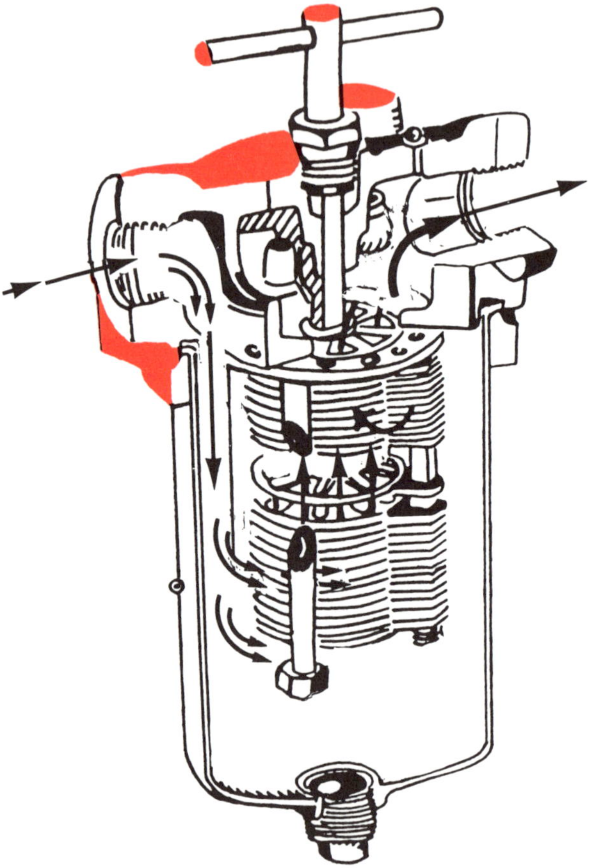

Figure 4-15. Cut-away view of disc cleaning filter.

Strainers are fitted with baskets that vary in diameter and mesh size to suit the particular grade of oil used. A fine mesh basket used with heavy oil will clog. A large mesh basket used with thin oil will not filter out impurities that can pass through and clog the nozzle orifice. The parts of a fuel oil strainer are as follows: hand screw; handle (showing which basket is closed and which is open); yoke; oil inlet; plug valve; drain; strainer basket; body of the strainer; spring handle; stuffing box gland; and locking flange.

When cleaning or replacing baskets, gaskets must be replaced carefully to ensure a tight seal and to prevent air from entering the oil line.

Fuel Oil Pump: The primary function of the oil pump is to move oil from a storage tank and deliver it to the nozzle of

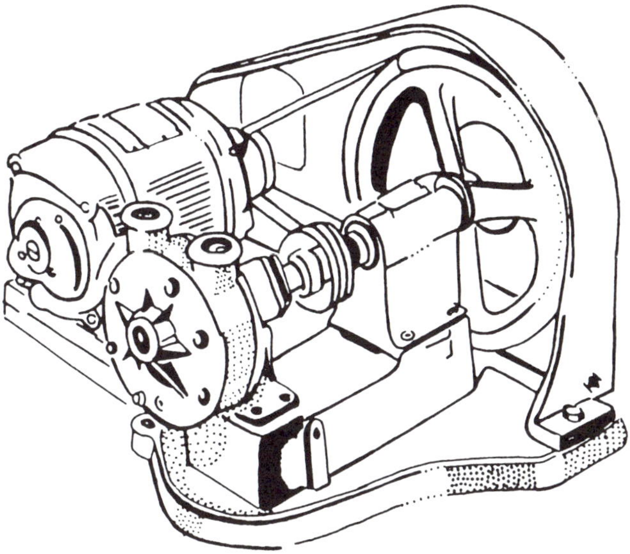

Figure 4-16. Fuel oil pump assembly.

the burner at a constant operating pressure. If the storage tank is located underground, the pump must perform double duty. It must not only deliver oil to the burner, but it must also lift the oil to the level of the nozzle.

Oil burner pumps are designed to deliver enough oil to exceed the nozzle output capacity so the pressure at the nozzle can be adequately controlled. Most pumps are equipped with a relief valve that discharges oil back to a return line to prevent excessive pressure build-up that may damage the pump or its packings.

Most fuel oil pumps are designed to run either clockwise or counter-clockwise to lend mounting flexibility. The pump assembly consists of the following: internal gear; idler gear; bearing and idler assembly; stuffing box; return seal; coupling; ball bearing; support for pump; V-belt drive; drive pulley; belt guard; and bed plate.

Fuel Oil Burners: Due to its fuel, an oil fired boiler has special requirements that makes it necessary to modify the design of its burners. As detailed previously, an oil burner must proportion and mix air and fuel. In addition, the burner must either vaporize oil or atomize it (place it into suspension) to prepare it for combustion. An oil burner must also constantly maintain a relatively high temperature around its flame. If not, impingement (the flame touches a cold surface) may occur and the resulting incomplete combustion will form accumulations of soot along with other harmful by-products.

Two basic types of oil burners exist, vaporizing and atomizing. Vaporizing burners have limited use in generating steam on any sizable scale due to the restriction of fuels they can burn. On the other hand, atomizing burners are capable of burning any fuel oil of just about any viscosity.

There are three practical methods for atomizing oil in preparation for combustion. Oil can be reduced to droplets by subjecting it to air or steam under pressure (which can be costly), by forcing oil (under pressure) through a nozzle, or by applying centrifugal force.

The most common burners are the mechanical atomizing types. One, the **rotary-cup oil burner** is used mainly with thicker oils because it can tear the oil into minuscule droplets. Oil is delivered at low pressure and discharged inside a brass cup rapidly revolving clockwise at 3600 rpm. The oil is spun around the cup and spread into a thin layer. The layer of oil moves towards the rim of the cup where centrifugal forces fling it off the rim and into an air stream spinning counter-clockwise. The counter-acting forces literally shred the oil droplets into small particles. The rotary-cup burner has gained wide popularity because no pressure is required for its operation other than to get the oil to the cup.

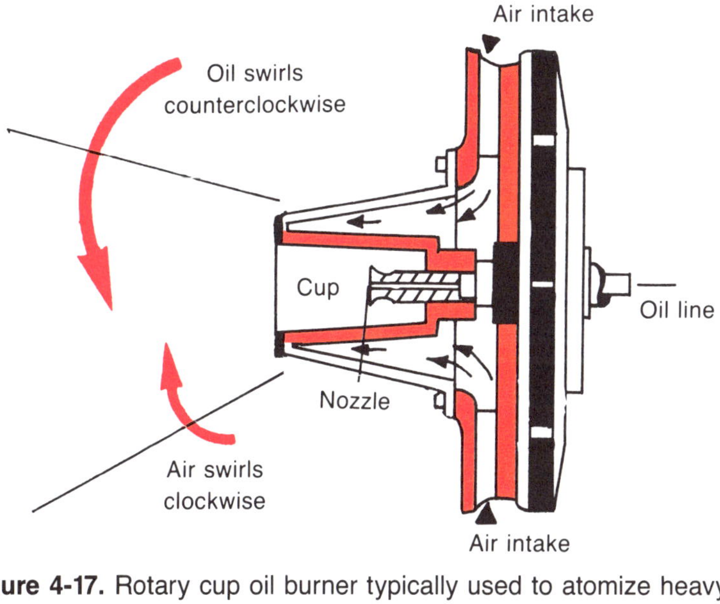

Figure 4-17. Rotary cup oil burner typically used to atomize heavy fuel oils.

A large majority of mechanical atomization burners rely on pressure alone to force oil through the small orifice of a nozzle so its dispersed into the combustion chamber as a fine spray or mist.

The **air atomizing burner** is an example of a nozzle equipped mechanical burner. Built into the front head of both high and low pressure boilers, its ability to produce a high degree of atomization with both light and heavy oils makes it very versatile. The air atomizer burner provides an intimate contact of oil and air that produces clean, complete combustion before flames can reach the furnace wall. This ability minimizes impingement and reduces secondary air requirements for complete combustion. Controls regulate both oil viscosity and pressure to the oil metering valve so the exact amount of oil needed to meet the boiler load demand is delivered to the burner.

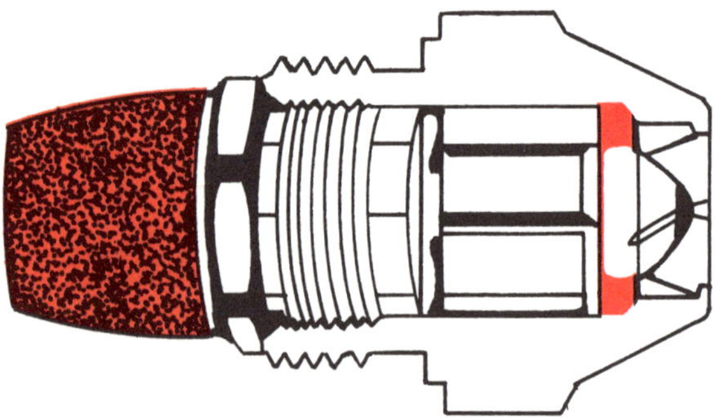

Figure 4-18. Sectional view of typical oil nozzle.

Designed to atomize oil and deliver a fixed amount of fuel to the combustion chamber, the nozzle is the primary element of the burner. It is expected to deliver atomized fuel

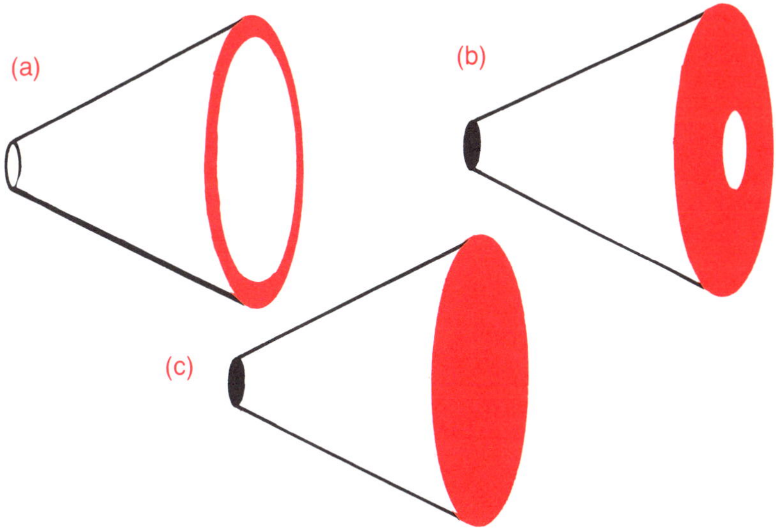

Figure 4-19. Nozzle spray patterns: a) hollow cone, b) all purpose cone, c) solid cone.

for combustion in a uniform spray pattern and at a spray angle best suited to the requirements of that particular burner. The spray patterns range from a hollow core cone to a solid cone to an all-purpose spray pattern.

Some nozzles are fitted with an adjustable control rod that can increase or decrease the size of the orifice. The control rod nozzle is particularly useful in applications where a wide range of oil viscosities and pressures are necessary.

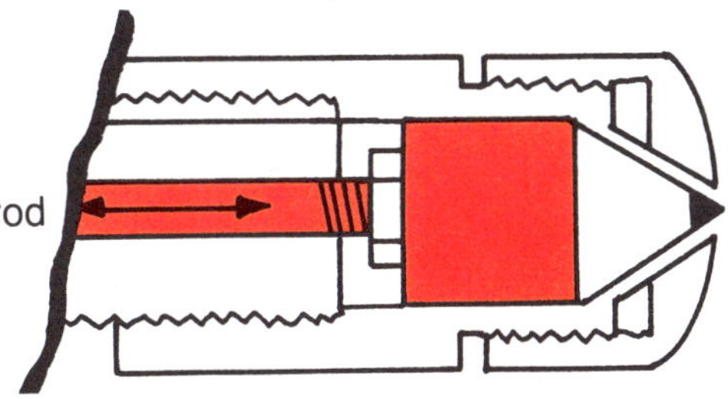

Figure 4-20. Cut-away view of adjustable nozzle.

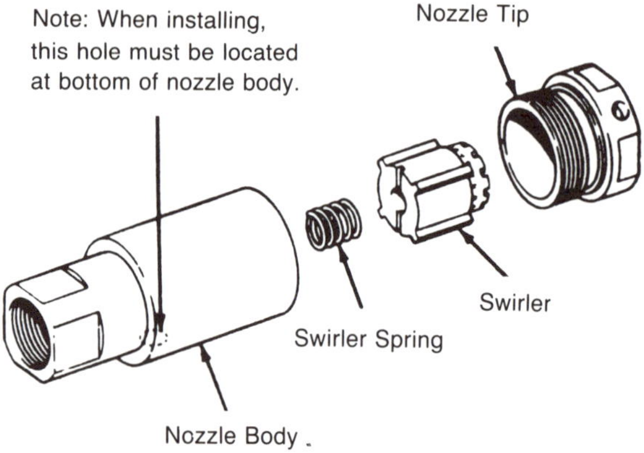

Figure 4-21. Break-away view of oil nozzle. *(Courtesy, Cleaver-Brooks)*

Nozzles are usually equipped with a strainer so dirt and other foreign material cannot clog the orifice and cause incomplete combustion. The nozzles of some air atomizing burners offer an advantage because they can be removed without opening or removing the front head of the boiler. Manufacturers often recommend changing the nozzle annually because the tips are subject to substantial corrosion by the oil. Unless the replacement meets the exact same specifications as the original nozzle design, the orifice size and the combustion chamber spray angle delivery may differ and adversely affect combustion and overall burner operation. The nozzle should be checked for wear periodically. If excessive, replacement may be necessary. If a burner is operating satisfactorily, many manufacturers advise against replacing the nozzle.

Many boilers have combination burners comprised of two burners, one oil and gas, connected to individual fuel lines. The burner can be switched to use either fuel simply by changing to the appropriate nozzle. For example, on the Cleaver Brooks boiler, this changeover from oil to gas is effected by lifting the safety interlock, removing the burner, clamping it to a clamp holder on the boiler and setting the other burner in its place. The safety interlock must be pushed back down into its slot to recouple the burner/fuel-line connection.

Due to economical considerations, triple service burners that burn oil, gas and pulverized fuel are also becoming increasingly more common.

COAL FUEL SYSTEMS

Coal is a widely used energy source. And because so many large steam generating plants are equipped to utilize coal, it will continue as a major energy source for quite some time. Even though coal is a relatively abundant fuel, the inherent problem lies in the fact that coal has a high sulfur content. When unrefined coal is used for combustion, sulfur escapes into the atmosphere and returns to the environment in the form of acid rain. The detrimental effects of acid rain upon the environment and the entire animal population, human and wildlife alike, are far reaching and devastating. Many scientific studies examining the wide spread effects of acid rain are underway. Even without the final conclusions tallied one fact is obvious. If use of coal as a major fuel is to continue, coal must be *cleaned* using some desulfurizing process to control pollution and protect the environment for future generations.

Generally speaking, the harder as coal is, the more fixed carbon it has and the less moisture and combustibles it contains. The softer the coal, the less carbon it has and the more hydrocarbons and moisture it contains.

Basically, coal can be divided into three main classes:

- **Anthracite** is the hardest coal, and it is primarily composed of carbon. Anthracite is characterized by having little volatile (combustible gaseous hydrocarbon) content and consequently ignites slowly. When burned, it requires a strong draft. The heating value of anthracite is approximately 14,000 Btus/pound.

- **Bituminous** is a soft coal that has a low carbon content and contains a higher percentage of combustibles. The heating value of bituminous is approximately 11,000 to 14,000 Btus/pound.

- **Lignite** is considered a low grade bituminous coal that only has an average heating value of about 9,000 to 11,000 Btus/pound. The grade of lignite varies so much that low grade lignite may only have 7,000 to 8,000 Btus/pound.

By considering the heating content of the various grades of coal, it becomes apparent that boilers equipped to use soft coal must be designed to have larger furnace volumes than those which burn hard coal.

Excess air is necessary to assure that all the combustible elements in the fuel receive adequate oxygen for complete combustion. For example, the excess air requirement for oil or gas fueled systems is 10% to 30%. Under good conditions, an average hand fired coal system requires 50% excess air. Under poor conditions, it is not unusual for the excess air requirement to approach 100%.

Over the years, coal firing has progressed from shoveling by hand to various mechanical stokers and pulverized firing. Coal may be burned on grates, on stokers or in suspension as pulverized particles.

Pulverized Coal Firing: Coal can be fired as a fine powder like fuel that is injected into a burner not unlike those use with oil. Coal ground into a fine granular powder is known as pulverized coal. By grinding up coal, the combustible fuel elements are exposed for quick ignition. As the coal particles come in contact with increasingly higher temperatures, the combustibles are drawn off as gas, mix with primary air and burn. As the volatile matter burns, secondary air enters at the burner and provides sufficient oxygen to complete the combustion of the remaining carbon particles.

Pulverized coal is usually blown into a furnace by a strong blast of hot air. The air used to blow the coal particles to the burner dries the coal and becomes the primary air for combustion. Pulverized coal burners are responsible for producing stable ignition.

Depending upon the type of firing employed at the burner, nearly all pulverized coal fired furnaces are completely water cooled due to considerable flame impingement. It is necessary to eliminate flames from blasting up against furnace walls because molten ash particles resulting from combustion of fixed carbon tend to adhere to and dissolve most furnace linings (refractories). Over an extended period of time, constant heat exposure and high sulfur contents gradually corrode the refractories.

Despite corrosive tendencies, suspension burning of pulverized particles is still the most effective method of coal firing and the most widely used in large boiler systems. Pulverized coal firing removes the limitation on the amount of fuel that can be burned in a boiler compared to stoker firing where a definite limit exists.

Coal which cannot be burned in suspension is burned on stokers.

Hand Firing: Uniform combustion cannot easily be maintained in a hand-fired grate furnace because the firing must be intermittent. Consequently, hand firing is basically reserved for small applications or emergency situations.

To hand fire coal, certain tools are necessary:

- Shovel—used to spread fresh fuel on the fire
- Slice bar—breaks up clinkers that stick to grate
- Hoe—used for ash removal on grate and ashpit
- Rake—used to spread the fuel evenly over grate
- Lazy bar—used to support the handles of the hoe and rake when cleaning the fire and ashpit

Spreading Method: A relatively thin, even layer of coal is manually spread over the grate. Coal is placed on the back of the grate and worked towards the front or the fire door. Due to the nature of hand stoking, small amounts of coal are fired at frequent intervals so the furnace temperature does not drop. Using a thick layer of coal only succeeds in wasting fuel and lowering the furnace temperature.

Alternate Method: By alternately firing each side of the grate, the combustible hydrocarbons released from the new layer of coal are quickly ignited by the fire on the opposite side of the grate. This method is preferred when soft coal (with a high volatile content) is fired.

Coking Method: A load of coal is fired at the front of the grate and allowed to coke. Once fired, the coke is spread over the grate in a thin layer and more coal is fired at the front.

Subsequently, hot gases derived from fresh coal at the front of the grate ignite. When the ignited gases pass over the fire of the burning bed of fuel towards the rear of the grate, combustion is completed.

Cleaning: Two methods are commonly employed to clean a hand fired furnace: cleaning alternate sides or front to rear. Hand fired boilers may have unmovable stationary grates or hinged dumping grates that are opened so ashes can be dumped into the ashpit.

In cleaning alternate sides, the top layer of coked fuel on the one side is swept over to the other side by using the hoe and slice bar. The clinker and ash are uncovered and are raked out with the hoe. The coked fuel is then swept back over the clean grate and fresh coal is fired. As the intensity of the fire is renewed, the opposite side of the grate is cleaned.

In front to rear cleaning, the coked fuel on the front of the grate is separated from the the clinkers and ashes with a hoe and worked towards the rear of the grate so the the ashes and clinkers can be removed through the fire door. Once they are removed, the coke is pulled forward and spread evenly across the grate.

When dumping grates, the good fire is pushed to the back, the front grates dumped and the good fire pulled forward. The front fire is built up, the back fire burnt down and the back then dumped. It is necessary to regularly clean fires so an even plant load can be maintained.

Mechanical Stokers: The most obvious advantage of using mechanical stokers over hand firing is the reduction of labor. More, fuel and firing conditions can be controlled more preciously to meet varying load demands.

The most common stokers are the sprinkler, traveling or chain grate, overfeed and underfeed stokers. The primary purpose of stokers is to evenly feed coal into the fire box while eliminating the need to open the fire door. By eliminating the need to constantly open and close the door to stoke the fire box, the furnace can maintain a more constant temperature and thoroughly burn the fuel. The basic components of mechanical stokers are the hopper, which holds coal, and a feed mechanism that stokes coal into the boiler. In addition, an integral fan provides air for combustion.

- Sprinkler Stoker: This type of stoker is adaptable and can be used with stationary grates, shaking grates or chain grates. This stoker design spreads coal over the grate replacing the fireman with his shovel. The stoker is usually mounted on the front of the boiler so it does not obstruct access to the furnace via the fire doors in case repairs and hand firing become necessary.

- Traveling or Chain Grate Stoker: In this conveyor type design, a coal hopper feeds fuel onto the front of a revolving grate which moves under a door that determines the thickness of the coal layer being carried into the furnace. As the bed of coal travels towards the rear of the grate, the speed of the conveyor is adjusted so the entire layer of fuel burns by the time it reaches the end of the conveyor. Once the bed of coal has been burned and it reaches the end of the conveyor, the ashes are dumped into a pit for removal by another conveyor.

- Overfeed Stoker: In this stoker, coal is transported from the hopper down a staircase like reciprocating grate. The coal on the grate steps is gradually consumed as it travels toward the bottom of the grate where the ashes are deposited on a dumping grate. The ashes on the grate are periodically dumped into an ash pit.

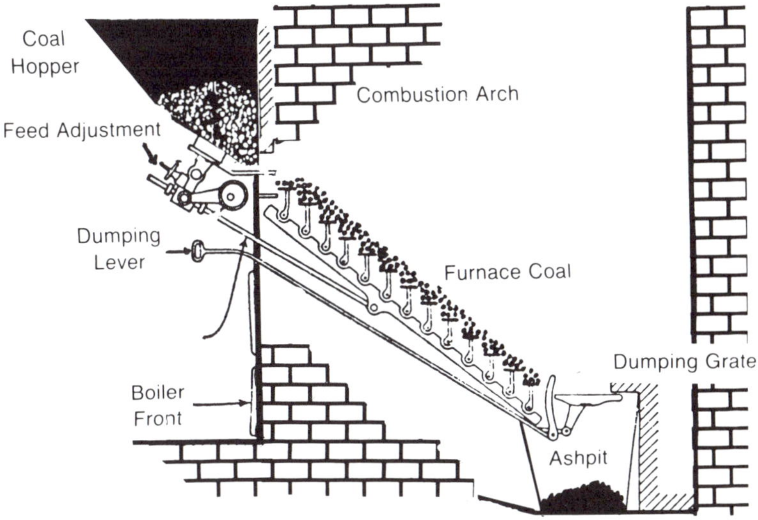

Figure 4-22. Overfeed mechanical stoker showing coal travel.

- Underfeed stoker: There are two main types of underfeed stokers: the screw feed and the ram feed. Both operate in much the same manner. Power driven rams push coal from the hopper into one or more pits (retorts). As the retorts fills with coal and finally overflows, inclined side grates pick up the overflow. As the remaining fuel is consumed on the grates, the ashes are slowly transported down the incline to dump plates. The dump plates are then tilted so the ashes are deposited into ashpits for removal.

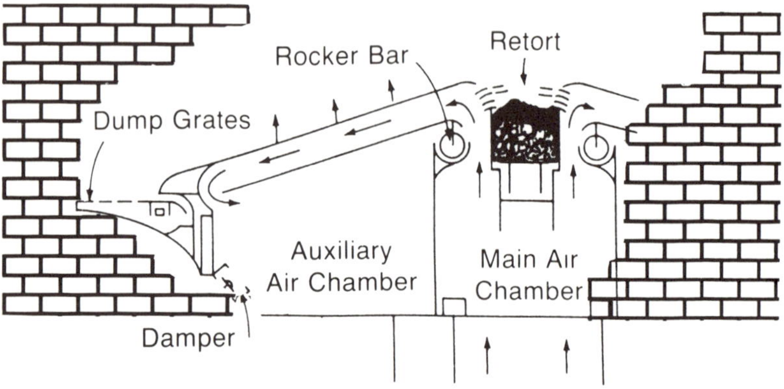

Figure 4-23. Mechanical single retort underfeed stoker.

In an alternate design, power driven rams push coal into a series of narrow retorts positioned just under the inclined grates and in between the grate steps. The coal overflows directly onto the grates. The grates usually provide a reciprocating action that helps move the coal downward to dump plates at the bottom of the incline where the ashes are deposited into an ashpit.

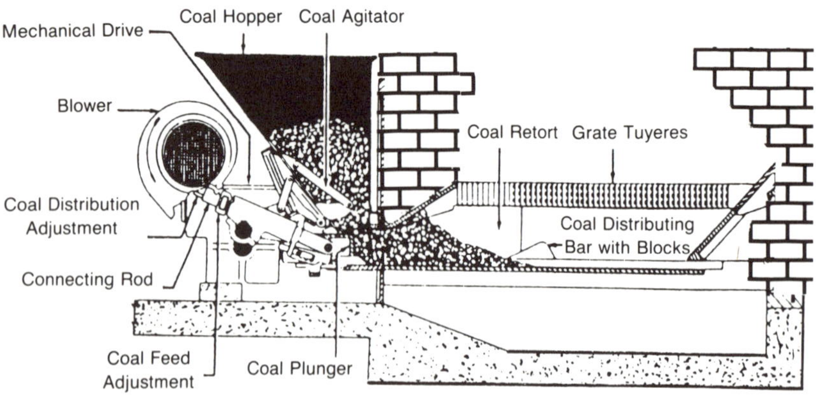

Figure 4-24. Alternate underfeed stoker design.

Many variations of underfeed stokers exist. In the common single retort design, a trough like retort traverses the entire length of the stoker and provides maximum underfeed combustion area. The opening of a large central retort expands to prevent packing of coal and to assure easy flow of coal onto grates. Located in the main air chamber, the retort is kept cool by air sweeping up from the lower portion of the air chamber. Natural, induced or forced draft may be used depending upon the size and design of the stoker. A coal feed pusher is also located within the retort and runs back and forth along the bottom of the retort to evenly distribute fuel over the entire grate area.

When starting stokers, the stoker should be greased and the oil level in the gear box checked. Fill the hopper and start running coal into the retort. The stoker is then disengaged until the grates are covered. Wood and waste rags are then placed on top of the coal for lighting. At no time is a volatile liquid to be used. The rags are then lighted and the outlet damper cracked open. Only a small amount of air should be running from the forced draft fan. When the coal starts burning, slowly start the stoker feed and increase the air. Warm the boiler up slowly.

When taking the boiler off-line, start early so that all the coal in the hopper is burned. When empty, the hopper is filled with ashes. The ashes will displace the coal in the retort, pushing out all the coal onto the grates to be burned. Gradually cut the forced draft so the firebox will slowly cool. After sufficient cooling, remove the ashes and coal from the retort and grates.

SECTION 4

FUELS AND COMBUSTION

Review Questions

1. What is combustion?
2. What three elements must be present to achieve combustion?
3. How does complete combustion differ from incomplete combustion? What assures complete combustion?
4. What products result from incomplete combustion and complete combustion?
5. How much air is required to sustain complete combustion? What results from using excess air?
6. What is the significance of primary air and how does it affect combustion?
7. Why must an operator strive to maintain complete combustion?
8. Identify the three most commonly used fuels.
9. What basic combustible element is present in all fuels? In which source does it occur as an elementary fuel in particle form?
10. Which is hotter, a clear (luminous) flame or a nonluminous flame? Why?
11. Define flash point and fire point.
12. To what stimulus do combustion controls react?
13. Identify the three basic types of combustion controls.
14. What do the combustion controls do if an unsafe condition occurs?
15. Which combustion control lends the best precision over a wide operating load range? Why?

16. Describe the sequence of the programming clock.
17. What device controls pilot and burner ignition to safeguard against explosions? What actuates it and what are the results?
18. How does a sulfide cell function within the programming clock?
19. What does flammability denote?
20. Fuel must be burned in suspension. What process is applied to oil and coal, respectively, to prepare for burning?
21. Why must fuels be reduced to small particulate matter prior to burning?
22. Why is turbulence necessary for burning?
23. Which fuels burn the easiest and liberate the most heat?
24. What element controls the consumption rate of fuel? What is the relationship?
25. How is city gas line pressure reduced?
26. What regulates the flow of gas into a boiler? What actuates this regulator?
27. Describe the function of the solenoid switch.
28. What determines the flow of gas to the burner?
29. Why do low pressure burners have a multi-jet configuration?
30. How is primary air introduced in a low pressure gas burner?
31. What provides secondary air for combustion?
32. High pressure gas burners require modified air/fuel mixers. Why?
33. Where does fuel become entrained with primary air in high pressure burners? Is secondary air required at the burner?
34. What ignites the air/fuel mixture?
35. What sensory device shuts off the gas main if the pilot flame is extinguished? How does it work?
36. Briefly describe the sequence used to light off a boiler, equipped with a gas burner.
37. What is meant by viscosity?
38. What is the relationship between viscosity and energy (Btu) content?

39. What are the four main grades of oil and which has the highest Btu content?
40. Why must some grades of oil be preheated? How is it accomplished?
41. Identify the two types of oil burners and list three methods used to prepare oil for combustion.
42. Describe how a rotary cup burner atomizes oil.
43. What is meant by a combination burner and which fuels can it burn?
44. What is the function of a nozzle? Where is it located?
45. What does an increase in a suction gage reading on a duplex strainer usually indicate?
46. Name the three main classes of coal. Which class has the highest Btu content?
47. Which has less carbon and a high hydrocarbon content, hard or soft coal?
48. List the three methods of firing coal. Which method has no limitation to the amount of fuel that can be burned?
49. Enumerate the three basic methods of hand firing.
50. What two methods are used to clean hand fired furnaces?
51. Identify the most commonly used mechanical stokers. What advantage do these offer over hand firing?
52. What are the two types of underfeed stokers? Briefly describe each.

5

BOILER OPERATION

SECTION 5

BOILER OPERATION

OPERATION PROCEDURES

Previous sections identified and explained the function of components and gave general descriptions of the various boilers operators are likely to encounter on the job. Now, actual boiler operation will be addressed.

When an operator begins his shift, standard procedures must be followed to assure safe and efficient boiler operation.

When coming on-duty, all log sheets from the previous shift need to be examined. Through readings and notes recorded by the previous operator, the log should reveal any malfunctions or discrepancies that may have occurred during the previous shift and whether or not all units and equipment are functioning properly. The log indicates which units are on-line (operational and running) and which are not. It also indicates what unit may need to be brought back on-line or what unit may need to be taken off-line later for maintenance or repair. Properly maintained logs also note water treatment records and should indicate whether any additional treatment has been prescribed to cure a particular condition.

Good log sheets lend valuable insight into what has occurred during previous shifts if some sort of diagnosis is warranted and must be made by operators on subsequent shifts. An alert operator can examine the log and may be able to detect potential operating problems before they happen. Consequently, the operator will be better prepared to take precautionary measures in order to head-off malfunctions or, at least, keep a problem from escalating to a potentially hazardous condition.

Check Out Procedure

Accurate and well documented log sheets are essential to any good plant operation. After reviewing the log sheets, an operator should proceed as follows:

1. Upon entering the boiler room, the first step is to check the water level on any boilers currently on-line. With steam boilers this is done by blowing down the gage glass. The water should enter quickly indicating clear lines which are free of sludge, sediment or scale build-up.

2. Next, blowdown the water column to make sure it is clear.

3. While the boiler is firing, blowdown the low water cut-off. The boiler should shut off. Provided the boiler is shut off, the cut-off should be closed, allowed to refill and the boiler should automatically light off.

4. Check the steam pressure and the fire condition for on-line boilers. If the flames are a very light, bright color with split ends, they may be getting too much air. If they

are hazy, puffy or smokey, the fuel mixture may be too rich or the burner may not be receiving sufficient air. If you are acquainted with the correct procedures, make the adjustments required. Otherwise notify another qualified operator.

5. Check the temperature, pressure and lubrication status on all running auxiliaries such as pumps, motors and fans. Verify that compressors and burners are not air bound.

6. While the boiler is firing, check the fire-eye scanner by blocking it to see if the boiler shuts off due to the simulated flame failure. Remember to clean the lens of the fire-eye after testing. Reset the programmer clock so it restarts at purge. Watch the firing cycle to be sure it lights off properly.

7. On oil fired boilers, shut down the burners and clean the burner nozzle or rotary cup.

8. The baskets of the oil strainer should be exchanged for clean ones. The baskets which were replaced should be cleaned and stored for backup use. In the event clogging or oil coagulating occurs and fouls the strainer, the backup basket can be switched with the fouled strainer baskets.

9. Be sure the feedwater system is working properly. If the boiler is taking on water more frequently than it should, either a leak in the system or a steam bound or blocked trap may be indicated.

10. Finally, make sure the safety valve has been properly checked and it is operational. If the safety valve has not been checked according to the schedule suggested by the manufacturer, check the safety valve or valves yourself. Also, make sure that all gages are reading accurately.

Shut-Down Procedure

When the need arises to take a boiler off-line, the following procedures should be followed:

1. Shut off and secure the fire.

2. After shutting off the fire, cut down on the draft to keep the refractory or brick work from cooling too rapidly. (Rapid cooling can cause flaking or spalling conditions.)

3. Watch the water level.

4. After a boiler has stopped steaming, close the main stop valve.

5. When the pressure drops to one or two pounds, open the air cock so no vacuum is formed. (Vacuum can cause tubes and shells to collapse.)

6. Do not close the main steam valve immediately or the safety valve may still pop open.

Start-up Procedure

If the occasion arises to put a boiler on-line, proceed as follows:

1. Check to verify that the water line is at the normal operating level.

2. Make sure main steam stop valve is closed.

3. If the boiler has been opened, repaired or inspected, check the fireside thoroughly to make sure tools or equipment were not accidentally left inside the boiler.

4. Purge the boiler to remove latent gases, excess oil or fumes that may remain in the fire box. Double check every combustion chamber with a visual inspection to verify that no accumulations of combusitibles are present.

5. Vent the boiler so no air is trapped in the steam space.

6. Set the boiler at the starting cycle and purge it again before firing.

7. Let the boiler warm up slowly.

8. Constantly monitor the water line to be certain it is maintained at the normal operating level.

9. When the steam gage shows pressure in the boiler, blowdown the gage glass, water column and low water cut-off.

10. Test the fire-eye to see if the boiler goes out on a simulated flame failure.

11. Watch closely until the boiler reaches two or three pounds below the header pressure. At that point, slowly open the main boiler stop valve until it is finally wide open. If opened too rapidly, the header can rapidly lose pressure. By gradually opening the valve, carry-over can be prevented.

Note: Never stand in front of a burner or boiler during start-up.

Blowdown Procedure

Periodic boiler blowdown is necessary for a number of reasons: to prevent accumulations of sludge and sediment, to control high water or high chemical concentrations and to dump or empty a boiler. An operator should always blowdown a boiler according to the following the guidelines:

Note: Only attempt blowdown at a light load.

1. Check the water level.

2. Slowly open the blowdown valve until partially opened. Once the purge is initiated, fully open the valve. Following draining, slowly close the valve.

3. Remain at the blowdown valve until the task is complete. Never leave a blowdown valve and walk away from the site.

4. Monitor the gage glass and do not blowdown to the extent that the water level drops to a dangerously low point.

INSPECTION

Boiler Inspections: Every boiler should be inspected at least once a year, regardless of whether the unit has displayed operational problems or not. Anytime a boiler has been subjected to substantial maintenance or a major repair, have the unit accessed by an authorized inspector prior to putting it back on-line. Before an operator tags a boiler for inspection, these procedures should be followed:

1. Take the boiler off-line and follow the procedure for securing a boiler which was outlined previously.

2. Close all boiler stop valves and tag the valves to warn other personnel that they are closed.

3. Ascertain that no vacuum exists in the boiler by opening a vent, air cock or try cock.

4. Close and tag the feedwater line. Make sure all pumps and motors are off.

5. As soon as the boiler has had sufficient time to cool, dump it and tag the boiler for inspection.

6. After dumping, remove the handholes, manhole covers and wash out plugs.
7. Immediately flush and wash out the waterside of the boiler so sludge and/or sediment accumulations cannot harden. If allowed to harden, cleaning may become difficult and costly. The fireside must also be thoroughly cleaned.

8. Check for signs of blisters, bags or cracks on the heating surfaces.

9. Remove the plugs from the water column and open the low water cut-off control to inspect the float chamber.

10. Schedule the inspection ahead of time so the boiler is not left open or unnecessarily taken off-line for any length of time.

Make it a practice to always have all the necessary equipment required readily available to expedite a thorough inspection. Before completely tearing down a boiler, determine what the inspector wants to focus on during the inspection. If the unit was inspected recently or the inspector is familiar with the boiler, only a partial inspection may be necessary.

Hydrostatic Test: There are times when an inspector may determine a hydrostatic test is warranted. If required, follow these procedures:

1. Completely fill the boiler with water, including the steam space on a steam generator.

2. If fitted with a whistle valve on the water column, remove it and plug the vacant access.

3. Close the main steam stop valve.

4. Safety valves have to be removed and replaced with blank flanges. Rather than removing the safety valves, they can sometimes be gagged with a clamp to prevent them from popping under pressure.

5. Keep the air cock open while filling the boiler. When water starts coming out of the cock, the boiler is completely filled and all air has been displaced. Once filled, securely close the air cock.

6. Bring the boiler pressure up to one and half times its maximum allowable working pressure (MAWP). This is usually accomplished by the boiler feed pump in combination with a air from a compressor capable of maintaining the required degree of pressure. Have an adequately sized compressor available if the inspector indicates this test is necessary.

7. Do not pump cold water into the boiler. Condensation or sweating may occur and be mistaken for a leak.
8. While the water pressure is maintained, check the boiler for leaks or other structural defects. Light hammer blows are occasionally used to test the strength of suspected weak parts.

9. After inspection, relieve the boiler pressure and drain the water to the normal operating level. Replace the blank flanges with safety valves or remove the clamps on the safety valves. Reinstall the whistle valve and any other parts that may have been temporarily removed for inspection such as the low water cut-off. Examine the handhole and manhole gaskets to verify whether they are safe to reuse. If in doubt, use new gaskets and replace the covers.

10. After testing, recheck all valves, safeties and working parts then put the boiler back on-line.

Boiler Lay-Up: When boilers must be off-line and out of service for any length of time, they should be laid up. Unless proper procedures are followed prior to taking boilers off-line, severe corrosion (especially by oxidizing agents) may occur. Oxygen from the atmosphere can enter an idle boiler, combine with condensed moisture and produce extensive pitting of metal surfaces which may compromise the soundness of the boiler.

The exact procedure employed to protect idle boilers depends upon the length of downtime. If downtime is anticipated to exceed one month, dry storage is preferred. Wet storage, on the other hand, is practical for shorter durations that do not exceed one month.

- **Wet Lay-Up**:

1. Remove all fuses from the burner circuit.
2. Remove soot and ash from the furnace, tubes and flue surfaces.
3. Drain the boiler completely.
4. Flush the boiler to remove all sludge and loose scale particles.
5. Repair or replace leaking tubes, nipples, staybolts, packing and insulation.
6. Clean and overhaul all boiler appurtenances such as safety valves, gage glasses and firing equipment. Pay special attention to low water cut-offs and feedwater regulators to make sure float (or electrode) chambers and connections are free of deposits.
7. Check the tightness of the condensate return system components to prevent leaks.

8. Seal the boiler, using new gaskets, and completely refill with treated, warm water to inhibit corrosion and oxygen pitting.

- **Dry Lay-Up**: Follow steps 1 through 7 above and thoroughly dry both the fireside and waterside of the boiler. Replace the handhole gaskets and handholes. Install trays of quick lime or silica gel on the waterside of the boiler to wick-up or absorb moisture from the air contained inside the boiler. Then replace the manhole gasket and cover. Close all connections and valves. The chemicals should be checked approximately every month and periodically replaced when saturated with moisture.

Gage Glass Cleaning and Replacing: When cleaning a gage glass, use a nylon brush from a rifle cleaning kit or a wooden dowel rod tipped with a cloth swab that will not scratch or break the glass.

If the glass gage leaks, it renders inaccurate water levels. Carefully tightening up the gage glass nuts may sometimes eliminate a leak. If not, the washers should be replaced. To replace a broken gage glass or worn washers, use the following procedure:

1. Secure the water valve to the gage glass.

2. Secure the steam valve to the gage glass.

3. Open the gage glass drain valve.

4. Remove the gage glass nuts.

5. Remove the broken glass and the washers.

6. Obtain and install the new gage glass and washers.

7. Take up on the gage glass nuts until hand tight and secure with 1/4 turn of a wrench.

10. Crack open the steam valve to the gage glass and let the glass slowly warm up.

11. Fully open the steam valve.

12. Open the water valve to the gage glass.

13. Close the gage glass drain valve and check for leaks.

14. Slowly warm the glass by gradually opening the system water valve.

Note: If the replacement gage glass must be cut to fit, allow 1/4-inch under the inside measurement so the glass can expand when it warms up. It is important to use new gage glass washers when replacing the glass.

CAUTION

High and Low Water

Boilers are similar to automobiles. No two are alike. You can have two which are the same model, each built in the same year, but both start and run differently, even though the same operating procedure is followed. Every piece of machinery has its own idiosyncrasy whether its a boiler or an automobile.

It is no wonder that two boilers, which for all intent purposes appear to be exactly alike, operate somewhat differently—even though they may have been manufactured on the same day and sport consecutive serial numbers. Experience alone teaches operators the quirks of how each boiler runs. For instance, the normal operating water level of every boiler may vary somewhat, despite the rule of thumb that states normal water level is the half-way increment on the gage glass. Under typical circumstances, if the water level drops to a point below the half-way mark of the glass, the boiler is developing a low water condition. If some water can be seen in the gage glass or water comes out of the bottom try cock, the water level is still high enough to safely allow water to be added.

Generally, if any water is visible in the gage glass, 2 to 3-inches of water still exists above the highest heating surface. Never attempt to add water to a boiler if there is no water visible in the gage glass. In this case, the risks of damaging the boiler or causing a boiler explosion are extremely high. Instead, protocol requires an operator to secure (shut down) the boiler, cool it down and notify the immediate supervisor and the boiler inspector. The secured boiler should then be inspected and the condition which caused the problem corrected. The same procedure should be followed when a fusible plug *drops* (melts down).

Remember, high water conditions can also cause serious problems. When water exceeds the normal operating level and mixes with the steam, carry-over occurs. The carry-over results in water hammer which weakens and may eventually rupture lines. Personnel who happen to be nearby when a line ruptures may sustain serious injuries.

Boiler and Furnace Explosions

The prevention of boiler or furnace explosions cannot be stressed strongly enough. Nearly eighty percent of all explosions have proven fatal to the on-duty operator.

Every explosion is investigated. Findings indicate that the majority are caused by human error. Due to strict construction codes and ridged safety requirements which boilers must meet or exceed, accidents can rarely be attributed to the boiler itself. Prior to shipment and installation, codes stipulate that packaged boilers be factory tested and approved under stringently controlled conditions that often exceed normal operation to safeguard against boiler failure.

The majority of furnace explosions, for example, are usually a result of an operator failing to purge combustible gases from the fire box before reignition is attempted. A high percentage of accidents caused by factors such as an undetected leaking fuel valve or an unacknowledged flame failure can also be traced back to operator failure.

If an operator notices any odor of gas around a particular boiler, it should immediately be secured. All gas lines, fittings and valves should meticulously be check for leakage. One of the oldest and yet most accurate methods to check for gas leaks is the soap bubble test. When soapy water is applied to suspect piping or connections, escaping gas is revealed by a series of small bubbles welling up from the leak. The bubble test can also be used to detect oil leaks. Leaking oil is usually indicated by a small, moist, oily residue or, in more excessive cases, as small spots or puddles.

To prevent explosions, operators must make it standard procedure to purge the furnace box and eliminate fumes or excess fuel that might be present before lighting. In cases where a leak is suspected, disconnect the igniter and reset the cycle on the program clock to be certain the purge phase is not bypassed before the reignition sequence is initiated. In the event a boiler needs to quickly be brought on-line or if a burner misfires at any time, **do not** turn the clock ahead to skip the purge cycle. Bypassing the purge mode may only succeed in building up an accumulation of raw fuel which, if left unpurged, may ignite and explode. Regrettably, and all too often, this sort of haste has cost human lives. Act responsibly and remember the axiom: the life you save may be your own.

SECTION 5

BOILER OPERATION

Review Questions

1. What is a boiler operator's primary responsibility when first coming on duty? How is this task accomplished?
2. Why must an operator examine the log sheets from the previous shift?
3. Enumerate the steps in the check out procedure in their proper order.
4. How is the fire-eye tested?
5. When checking the fire condition, too much air results in what kind of flame?
6. If a leak develops in the system or a steam trap becomes blocked, what condition occurs?
7. Briefly describe the procedure for taking a boiler off-line.
8. What causes tubes to collapse?
9. List the steps taken to put a boiler on-line.
10. Why are boilers blowndown?
11. Detail the blowdown procedure.
12. When should a boiler be inspected?
13. Why must a boiler be flushed and washed out after dumping?
14. Under what conditions is a partial inspection necessary?
15. What is a hydrostatic test? What boiler pressure is required to conduct this test?
16. Why shouldn't cold water be pumped into a boiler for hydrostatic testing?
17. What procedure is used to protect idle boilers? Why?

18. If a boiler must be taken off-line for a period longer than one month, what should be done? What materials are used to absorb moisture? To be effective, how often must these be replaced?
19. When replacing a gage glass, the new glass must be undercut by a 1/4" increment to allow for ___________.
20. What should be replaced when installing a new gage glass to eliminate leakage?
21. If any water is visible in the gage glass, generally how much much system water does a boiler contain?
22. The half-way increment on the gage glass indicates what?
23. Complete the statement: It is still safe to add water to a boiler as long as...
24. What problems can high water conditions cause?
25. What is the most prominent cause of boiler explosions?
26. Explosion investigations usually reveal that what element can be credited with the accident?
27. What is the oldest and yet most accurate test to reveal fuel leaks?
28. What sort of leaks can be detected using that test?
29. What is the most important step an operator can take to prevent boiler or furnace explosions?
30. Before attempting relighting, an operator must never...

6

WATER TREATMENT

SECTION 6

WATER TREATMENT

WATERSIDE PROBLEMS

The waterside surfaces of boilers are continually plagued with problems such as scale, corrosion, fouling and foaming. Scale deposits on heating surfaces increase boiler temperatures and lower operating efficiency. Corrosion damages metal surfaces which can lead to metal fatigue and failure. Fouling clogs circulation. Foaming results in overheating and causes carry-over which can transport water impurities to the steam system.

Water Impurities

The idea that if water is pure enough to drink, it is pure enough for any use is a commonly held misconception. City water, wells and springs all contain minerals and impurities, even though they are not visible.

Water, by its very nature, is a solvent. Any water used in heating or steam systems is actually a solution, a combination of water and the impurities it contains.

Water picks up impurities from the environment. The type of impurities water contains depends a great deal upon the water source. The amount of those impurities varies according to the length of time the water remains in contact with the atmosphere. Even rain water accumulates sufficient oxygen and carbon dioxide from the atmosphere to be corrosive.

Impurities can be divided into three distinct categories: dissolved mineral salts, dissolved gases and suspended solids. Any one or a combination of these impurities lend water undesirable characteristics.

If water is to be used in a boiler, special treatment is required to eliminate or minimize problems which can be caused by dissolved mineral salts and gases.

Deposits

Minerals salts dissolved in water are the principle source of deposits in boilers. Scale is the most prevalent. Scale deposits adhere to waterside surfaces and cause corrosion. This is especially true in steam generators where fresh water is constantly added, heated and converted into steam. After water has been vaporized and the resulting steam separates from the water, minerals remain in the water and have the potential to precipitate out as scale.

Minerals, in the form of scale or sludge, which are deposited on heating surfaces impair the boiler's heating or cooling capabilities. The deposits tend to insulate heating surfaces which subsequently reduces heat transfer and boiler efficiency.

When heat transfer decreases, energy consumption increases and operating costs rise.

If a scaling condition becomes excessive in firetube boilers, tubes may bulge due to overheating. Consequently, tubes can burn out, or in more extreme cases, burst. Scale has also been known to clog the inside of tubes on watertube boilers. Clogged tubes impair circulation and raise boiler pressure, increasing the potential of an explosion.

Sludge is a deposit formed by the coagulation of salts and other solids present in boiler water. Sludge appears as soft lumps or thick masses of material that settle out in the low points of a boiler and form a layer of "mud."

Unlike scale, sludge can be purged from a boiler by blowdowns or by thorough washings. If not controlled, sludge can clog boiler circulation and eventually bake out or harden. Light sludge build-up can insulate tubes and hamper circulation. In extreme cases, especially with baked out sludge, overheating and high operating pressures can result.

Corrosion

Briefly defined, corrosion is a condition evidenced by the deterioration or wasting away of metal surfaces within a boiler. Corrosion occurs in many different forms including: pitting, grooving and caustic cracking (embrittlement). It can occur both internally or externally.

External corrosion is most often exhibited as rust or oxidation but can also occur in firetubes and flues due to the caustic affects of sulfuric acid.

A considerable amount of water is produced during the combustion process. Under ideal conditions, this moisture is carried to the flue and vented to the outside. Depending on operating conditions of a boiler, the moisture being carried by the combustion gases can condensate in the firetubes, flue or chimney or can be absorbed by soot accumulations that build up in improperly cleaned boilers. Recall that incomplete combustion produces flue gases that usually contain sulfur dioxide and other undesirable by-products such as soot and aldehydes. The sulfur gas resulting from incomplete combustion (or sulfur contained in soot accumulations) can combine with the condensate or moisture entrained in the combustion gases to from sulfuric acid. Sulfuric acid can corrode both metal and masonry surfaces.

Internal corrosion can be attributed to acids, oxygen (and other gases present in feedwater) or by electrolysis.

- Galvanic corrosion is due to electrolytic action which is the transfer of molecules from one metal to another dissimilar metal in the presence of water.
- Uniform corrosion, which is usually formed by acid, is very common. It is the easiest to control with water treatments.
- Pitting corrosion is accelerated localized deterioration commonly caused by a flake of scale reacting with oxygen dissolved in water. Though deep pitting occurs, it has little effect on the surrounding metal.
- Grooving is a uniform corrosion evidenced by narrow groves which often appears along the edges of riveted joints.
- Caustic cracking or embrittlement is a crystallizing or hardening process that causes fine cracks in metal, especially around riveted joints or holes. This type of corrosion is particularly dangerous because it weakens the metal structures of the boiler such as shell walls and joints which are then more susceptible to stress under high operating pressures.

Foaming

System water with a high concentration of impurities may encounter a problem called foaming. This condition is characterized by an oily scum on the water surface that traps and retains unbroken steam bubbles preventing steam separation. The principle problem with foaming is that impurities become entrained in the steam and are then carried over into the steam system. Once impurities have worked their way into the steam system, they can form deposits which damage heaters and other system components.

Foaming can be controlled by filtration, antifoaming agents and treatment to remove solid impurities. Surface blowdown can also help minimize foaming.

Fouling

Another problem experienced in boiler operation is fouling. Fouling can be defined as the accumulation of solid materials that either hampers boilers by clogging nozzles and pipes, which restricts circulation, or otherwise reduces heat transfer efficiency. Fouling can also contribute to deterioration. Common foulants are: silt, sand, grease, bacteria, algae, oil and corrosion products. Most of the time, fouling (with the exception of scale and corrosion products) is a problem that can be solved mechanically with filters. Water foulants should be filtered out of water before introduced into boiler systems.

WATER TREATMENT

Terminology

A review of some of the terminology applicable to treating water in order to counteract corrosion problems follows as a preface for addressing water treatment.

Parts per million (ppm) Ppm measures the amount of any elements which may be present in water. When testing water, it is used to specify exactly how much of a certain material or element (such as water treatment chemical) is dissolved in the water.

Grains per gallon (gpg) Similar to ppm, this is an alternate unit of measurement that specifies the amount of material dissolved in the water. For conversion, gpg multiplied by 17.1 equals ppm.

Concentration This term is used to indicate the strength or relative amount of an element present in a solution such a feedwater. While high concentrations of soluble impurities may not form scale, other problems can occur such as foaming, carry-over and corrosion. Concentrations of soluble impurities and salts contained in boiler water can be controlled or reduced by blowdowns.

Scale Scale is a dense coating of mineral matter which precipitates and settles out on internal heating surfaces as a result of falling or cooling water temperatures.

pH A value used to express the natural tendency of water to either form scale or corrode metal. The pH scale runs from 0 to 14. The mid-range increment, 7, indicates the neutral point. Distilled water, for example, has a pH of 7 and is neither scale forming nor corrosive. Any reading below a pH of 7 indicates the water is acidic and corrosive. Readings above 7 indicate the water is alkaline and potentially scale forming.

Alkalinity This factor is a measure of the ability of water to neutralize acid. The amount of alkaline material present in the water is directly related to pH. Usually, as alkalinity increases, the solubility of calcium carbonate (the most predominant scale constituent) decreases and it precipitates out as scale.

Like pH, alkalinity can be used as a test or control point for water treatment. The intensity of alkalinity is measured as ppm or gpg. Alkalinity has proven to be the most accurate indicator of whether water will foster a scaling condition or be corrosive.

Solids Solids exist in water either in a dissolved state or as suspended particulate matter. Suspended solids can be filtered out of boiler water; dissolved solids cannot.

Dissolved solids are predominately calcium, magnesium, chloride and sulfate. Nitrates, silicates and organic material from sewage, industrial waste or substances leeched from natural sources are often present in water as well. Dissolved solids intensify the corrosive effect of oxygen and carbon dioxide. The higher the concentration, the greater the water conductivity. As the water conductivity rises, the potential for corrosion increases. (When present in sufficient quantities, any dissolved solid can react or break down to form scale.)

The term total solids constitutes both dissolved and suspended solids collectively. Total solids are measured and pressed in ppm or gpg. As with hardness and chlorides, by monitoring solids blowdown maintenance can be scheduled to lower system water concentrations.

Cycles of Concentration Cycles are determined by the ratio of dissolved solids in system water to dissolved solids in makeup water.

When water evaporates, the dissolved solids it contains do not travel with the steam (evaporated water). Rather, they remain in the water. Boiling water to generate steam increases the amount (concentration) of solids in the water. As fresh water is added to the existing water, and then boiled off, additional solids are left behind raising the concentration even more.

In water treatment, cycles indicate when the concentration of an element in boiler system water has risen above the concentration contained in the makeup or feedwater. For instance, if the hardness in boiler water is determined to be three times as great as the hardness in the makeup or city water, then the boiler system water would have three cycles of hardness.

Hardness Considering its low solubility, the presence of calcium (limestone) in water is a primary concern because nearly 90% of scale deposits are composed of calcium carbonate. The hardness of a water sample constitutes the amount of dissolved calcium salts and/or magnesium present. Hardness is measured in either ppm or gpg. Testing for hardness can be used to monitor salt and impurity concentrations in boiler water.

Chlorides Chlorides are the soluble mineral salts that remain after water has been boiled away. As water becomes concentrated, chlorides are one element that do not react with other dissolved or suspended minerals. When makeup water is added to a boiler or hot water system, the chloride concentration increases until the water can no longer hold it in suspension. The salts precipitate out and cause corrosion.

Since a high chloride concentration makes water corrosive, system water is periodically tested for chlorides to determine how much water should be drained during blowdown to properly dilute the concentration. By monitoring chloride concentrations, preventative measures can be taken to minimize scale formation.

By comparing chlorides in makeup or feedwater with the

chlorides in the system water, the degree of chloride concentration in the system water can be determined (cycles of concentration). The concentration of soluble chloride salts present in water is measured in ppm or gpg.

Treatment Any chemical or material added to the water to produce some type of action that controls corrosion or scaling is considered water treatment. Treatment materials can either be liquid, powder or pellets. Any water treatment must be monitored by testing to verify that the proper doses have been administered and that the expected results have been attained.

Inhibitor Any water treatment used to protect metal against corrosion is termed an inhibitor. Inhibitors coat metal with a protective film that prevents electrochemical corrosion and oxygen attack. Nearly all closed loop systems, especially hot water heating systems, use inhibitors. Many types of inhibitors provide good results, including: chromate, nitrate, zinc and polyphosphates.

Blowdown Water cannot be indefinitely added to boiler system water without causing a high concentration of impurities. The concentration of total solids is controlled by blowdown (expelling system water) at prescribed intervals.

Oxidizing Agents These are compounds that are either oxygen rich and can readily release oxygen molecules or are compounds that attract electrons during a chemical reaction. One of most commonly encountered types of corrosion stems from dissolved oxygen coming in contact with metal surfaces and causing pitting. The areas most often affected are the feedwater heater, feedwater lines and the internal steam drum surface.

Temperature Corrosion rates are influenced by temperature. By increasing temperature, chemical reactions take place more rapidly. System water at temperatures above and below the 160° to 180°F range demonstrate lower corrosion rates.

Water Velocity Corrosion rates are related to the speed with which water circulates. High velocities increase the amount of oxygen in the water, and subsequently increase corrosion. Water moving at high velocity can erode metal surfaces in pipes as well as the protective films laid down by inhibitors. High water velocity also interferes with the initial formation of inhibitor films. Low velocity, on the other hand, can increase corrosion because suspended solids have a chance to accumulate more easily. The deposits can then cause pitting.

WATER TREATMENT METHODS

Chemical Treatment

Hot Water Heating Systems: Hot water heating systems are closed loop systems. The primary concerns with closed loop systems are to prevent air from seeping in and water from leaking out.

Since hot water boilers do not require makeup water to supplement system water, water treatment is much less complex than that required in systems where makeup water is a constant necessity.

Contained in both air and water, oxygen is one of the main agents of corrosion. To combat oxygen related corrosion, the proper alkalinity level must be maintained through the use of inhibitor treatment.

Water temperature within a boiler must be maintained above 180°F to eliminate the possibility of corrosion on the fireside of the boiler tubes. If the water temperature falls below 180°, the water vapor may condense within the fire-tubes. Then combustion gases, which contain carbon dioxide, can combine with the condensate to form carbonic acid. The resulting formation of carbonic acid can then corrode the metal tubes from within.

Steam Generators: The system water in steam boilers must continually be monitored and treated to prevent corrosion. It also requires treatment to prevent scaling. Due to steam and water losses, some quantity of makeup is often required in steam boilers. Since makeup water is used, periodic blow-downs are necessary to control chloride and hardness levels.

Treatment for scale prevention allows higher cycles of concentrations but inhibits scale formation. Treated with tannin or lignin, solids in system water precipitate and settle out in mud drums as sludge, which can be purged through blow-down. Often added along with slugging compounds, phosphates inhibit corrosion and also allow an increase in the cycles.

Alkaline water treatments such as soda ash, caustic soda (lye), sodium silicate or sodium phosphates tend to raise the pH of water. To minimize corrosion, boiler water pH is routinely increased and maintained between 10.5 and 11.5. Although high pH levels prevent corrosion, they also promote scale formation. Consequently, in order to control scale formation, inhibitors are always used in conjunction with escalated pH levels.

Corrosion sometimes poses a problem in steam condensation return lines. If corrosion is detected, due to the presence of either oxygen or carbon dioxide gas, a treatment such as sulphite can be utilized to eliminate oxygen and arrest its corrosive action. Sulphite combines with oxygen to form sulphate, an element which is not harmful to the system. In addition, treatments such as hydrazine or filming amine can also be used to line the interior surface of the return pipes with a protective film. The film protects the surfaces against carbonic acid attack which results when water and carbon dioxide combine.

A principle factor that contributes to corrosion is the presence of noncondensable gases (i.e., oxygen, carbon dioxide) especially in feedwater. Basically, oxygen contained in feedwater is liberated during steam generation. Still, a residual amount of oxygen remains that can corrode metal. Oxygen attack or oxidation is a type of corrosion evidenced by pitting of metal surfaces. Carbon dioxide has the potential to recombine and form carbonic acid which predominately causes corrosion inside horizontal pipes. The acid etches narrow trenches into the surface where the condensate travels.

Due to the corrosive effects of noncondensable gases in water, it is necessary to deaerate the water either with chemicals or by mechanical means (deaerating heater). Oxygen can be removed from water chemically by the use of oxygen scavengers. Both sodium sulfite and hydrazine are effective in this application. Continuous chemical feed and careful control of residuals are necessary to assure best results in oxygen removal.

Strictly adhere to the instructions issued for the water treatment program in your plant. No *standard* treatment exists. Chemical dosages and treatment levels vary according to the system designs. Each case is unique and must be diagnosed and treated accordingly. Some water treatments are composed of several different chemicals blended together to effectively control several conditions simultaneously.

Remember, minerals that exist in water cannot be destroyed. With the proper chemicals, the properties of those minerals can be altered to make them more easily manageable. For instance, scale forming salts and minerals can be changed into non-adhering sludge. The advantage of transforming salts into sludge is that it can be blowdown and purged from the boiler. In order for sludge to settle out, the water in a boiler has to be on a light load to slow the circulation or boiling process. Once settled, it can be expelled by blowdown.

Priming, Carry-over and Foaming

Other water conditions that require treatment include: priming, carry-over and foaming. These conditions can be attributed to high water levels, high chemical concentrations, water impurities that yield a high surface tension or by opening a main steam valve too quickly.

A boiler must generate clean steam to operate efficiently and economically. In carry-over, impurities entrained in otherwise clean steam are carried over into the steam lines, sometimes along with slugs of water. Priming, on the other hand, is the transference of water (either as a spray or as slugs) into the steam lines. Both priming and carry-over cause what is known as water hammer. Water hammer vibrates steam lines and weakens joints. If neglected, lines may rupture, especially at a 90 degree turns.

The impurities in steam due to carry-over can be eliminated by steam washers or separators. Washers use clean feedwater to literally wash steam; separators remove water and impurities from steam by routing the steam through a series of baffles so the flow of steam changes direction and the solids are separated out. Separators also help eliminate *slugs* of water that are carried over and may damage steam lines.

When boiler water has a high surface tension (due to a high concentration of solids), steam bubbles become encased by a film of water. These encased bubbles rise and fill the steam space. Foaming occurs when the steam space is partially filled with unbroken steam bubbles. This condition is typically caused either by water impurities that prevent the clean separation steam as it rises to the surface or by an oily scum on the water surface. Foaming is controlled through treatment and filtration of system water to remove the solid impurities. Surface blowdown is also valuable in ridding the water of oil scum. Surface blowdown, performed on a regular basis, not only reduces impurities and minimizes heating surface deposits, but it also reduces heat and chemical loss.

Mechanical Water Treatment

Evaporators: Commonly used along feedwater lines, these units are used to alleviate high concentrations of solids that can result in foaming conditions. By applying heat and vaporizing water, evaporators effectively remove both soluble and suspended insoluble solids from makeup water. The solids are retained by the evaporator and the vaporized makeup water is condensed and finally routed to the boiler, free of solid particulate matter.

Softeners: Hard water can be softened by utilizing a softener. Many of these units use zeolite, a sand like substance which is arranged in a filter bed. Calcium and magnesium in hard water filter through the zeolite bed and are replaced by

Figure 6-1. Packaged water softener assembly. *(Courtesy, Cleaver-Brooks)*

sodium (salt) derived from the zeolite. Calcium bicarbonate, for example, becomes sodium bicarbonate; magnesium bicarbonate becomes a sodium bicarbonate. Since sodium compounds are not scale forming, the conversion softens the water.

Through extended use, the sodium content of the zeolite bed is gradually depleted. Once the sodium supply has been exhausted, the unit can no longer soften water and the mineral salts pass through without being affected. However, the unit's ability to soften water can be restored merely by infusing the unit with a dose of strong brine composed of common table salt (sodium chloride). The brine solution rejuvenates the zeolite bed and *recharges* the softening unit.

Deaerators: Deaerating heaters are very effective in removing noncondensable gases from feedwater. These units are often used in conjunction with chemical treatment to rid water of highly corrosive oxidizing agents. Since oxygen becomes less soluble as water temperature increases, removing the oxygen is easily accomplished by heating feedwater to its boiling point. As oxygen and other gases are driven off and vented from the heater, other impurities (which cause hardness) precipitate and settle out to the bottom of the heater. Most deaerating heaters are equipped with filtration systems that contain the solidified impurities rather than allowing them pass on into the system water.

In a deareator, feedwater is discharged as a fine spray into the upper portion of a closed chamber. Steam enters and mingles with the finely divided streams of water subjecting them to a violent scrubbing action. The cleansing process continues as the water trickles down over bands of small trays while another traversing stream heats the water to liberate the

Figure 6-2. Packaged deaerator unit. *(Courtesy, Cleaver-Brooks)*

noncondensable gases. After coming in contact with the feedwater, most of the steam condenses. The remaining vapors, along with the separated gases, enter a vent condenser where steam condenses as it comes in contact with tubes that contain cool incoming water. The gases are vented to the atmosphere. If the system is operated below atmospheric pressure, the gases are drawn off by a vacuum pump and vented.

- In an **open heater**, the feedwater is heated until it approaches its boiling point. The heat causes the carbonates of lime and magnesium to solidify. The noncondensable gases (oxygen and carbon dioxide) are driven out of solution and vented to the atmosphere. Open heaters usually have filter compartments to trap the precipitated solids and other feedwater impurities.

- In a **closed heater**, the temperature is sufficiently increased to make the sulfates and carbonates of lime and magnesium solidify and precipitate out of solution. Sulfates remain in solution in temperatures up to 300°F, after which the solidify and settle. Carbonate, on the other hand, precipitates out at a much lower temperature.

Since impurities tend to produce scale deposits on heating surfaces, water should be chemically treated prior to introducing it into a closed heater. A closed heater consists of a steel shell that houses a large number of small tubes that contain and circulate steam. The water and steam do not come in contact with each other.

A closed feedwater heater is usually placed between the feed pump and a boiler. This arrangement allows water passing through the heater to have a greater pressure than the system water in the boiler. Due to the increased pressure, higher feedwater temperatures can be maintained. In contrast, if the water temperature in an open heater is increased proportionately, it would flash to steam. Many systems have an open heater installed before the pump and a closed heater after the feed pump.

Oil Heaters: Because heavier oils (Bunker C or #6 fuel oil) must be heated by boiler water or steam before being pumped to the burner, oil leaks are an ever present problem. Escaping oil tends to settle and adhere to heating surfaces almost immediately. The result, overheating. Tubes may show signs of oil fouling by bulging, sagging or burning out. Remember, oil leaks have the potential of developing into extremely dangerous conditions. Boiler explosions due to oil fouling are not uncommon. If an oil leak is discovered, immediately shut the boiler down.

Once oil leaks into system water and coats tubes, it is nearly impossible to remove. Often, if a leak occurs and goes undetected for even a short time, the boiler may have to be retubed.

If a very small quantity of oil works its way into system water, it increases the surface tension of the water and causes foaming. In some instances, minor oil fouling can be cured by boiling out the oil with an alkaline material such as caustic soda or soda ash. A thorough cleaning must follow.

Note: Condensate derived from heaters, which may be contaminated with oil, should not be returned to condensate tanks for recycling.

Cooling Towers: Large boiler installations often utilize water cooling towers. Open type towers are highly susceptible to algae growth if located where the sun can beat down on the water surface. Algae is one common source of organic fouling.

Treatment with chemical algaecides can prevent or slow down algae growth. Still, many times algae growth is so dense, it must be manually scraped off tower surfaces so surface openings or holes are not blocked.

Conclusion

Proper water treatment is an important aspect of boiler operation. It not only saves equipment, it also minimizes maintenance costs, saves fuel and electricity and prevents expensive repairs.

SECTION 6

WATER TREATMENT

Review Questions

1. What are the four maladies of waterside surfaces?
2. By its nature, what is water?
3. List the three categories of impurities.
4. Boiler deposits are primarily composed of what?
5. What kind of deposit is most prevalent? Why?
6. How does this deposit affect heating surfaces?
7. When watertubes become clogged by deposits, what results?
8. What is sludge? How is it removed from a boiler?
9. Define corrosion. List the forms of internal corrosion.
10. How is external corrosion most often exhibited?
11. Moisture produced during combustion sometimes combines with what to form caustic sulfuric acid?
12. What condition is characterized by an oily scum on the water surface? What causes it and how can it be controlled?
13. How can fouling be controlled?
14. List the common foulants.
15. What does the pH value indicate? What is the neutral point on the pH scale?
16. What does the abbreviation ppm express?
17. What is alkalinity?
18. What element is the most predominant scale constituent?
19. What does a hardness check indicate?
20. What term is used to indicate mineral concentration present in water?

21. Define cycles of concentration.
22. What intensifies the corrosive effect of oxygen and carbon dioxide?
23. How can chlorides be controlled?
24. What is an inhibitor?
25. What components are most commonly affected by oxidizing agents?
26. State the relationship between temperature and corrosion. Between water velocity and corrosion.
27. What is the result of a high pH level?
28. What is used in combination with high pH levels?
29. What elements are used to raise the pH in system water?
30. Which element can be used to eliminate oxygen and its corrosive action? What results?
31. Identify the chemicals used to eliminate oxygen from water.
32. Describe the standard water treatment program.
33. Define carry-over and priming.
34. What results from high water surface tension?
35. What does surface blowdown accomplish?
36. Besides using chemicals, how can water treatment be accomplished?
37. How can a zeolite softener be rejuvenated?
38. What function do deaerators perform?
39. How do water temperatures affect oxygen solubility?
40. Briefly describe how a typical deaerator operates.
41. What should be done if an oil preheater develops a leak? What condition may result?
42. If a leak goes undetected, what may have to be done? Why?
43. Should condensate derived from fuel oil heaters be returned to the condensate tank? Why or why not?

AFTERWORD

Under the direction of proficiently trained operators, there is no question that boiler failures, whether due to mechanical or human error, can be minimized. No substitute exists for astute, conscientious boiler operators educated in proper operating procedures.

It is imperative that operators set up a scheduled routine to test and maintain controls and safety devices. Be aware that any control or safety device is only as good as the personnel who have been assigned to test and maintain it. Even the most modern and sophisticated controls and safety devices cannot guarantee that a boiler will not malfunction. The fact of the matter is that only the human element, the operator, can guarantee safe and efficient boiler operation.

APPENDICES

APPENDIX A

BOILER ROOM SAFETY

On the job safety must be a primary consideration for any boiler operator. Safety regulations in both boiler and equipment rooms are crucial in providing safe work conditions for boiler operators. While on-duty, responsible operators must know and strictly adhere to the following list of safety precautions which was derived from the *National Association of Power Engineer's Magazine*.

SAFETY PRECAUTIONS

1. **Never** wear frayed clothing that might catch in machinery or cause you to trip.

2. **Never** wear shoes with worn soles that can easily be punctured or which can cause you to fall on slippery surfaces.

3. **Always** wear a hard hat or protective helmet where the danger of head injuries due to electric wiring or falling objects may exist.

4. **Always** protect your eyes from flying particles or the extended ray of a welding machine with appropriate goggles.

5. **Always** protect your hands from the dangers of burns, slivers and abrasions.

6. **Never** enter a remote area of a large vessel or boiler without a dependable light source.

7. **Always** use a flashlight that contains fresh batteries and provides ample light.

8. **Never** use an extension light with a weak or frayed cord or one without a cage to protect the bulb.

9. **Never** tamper with electrical circuitry unless you have had formal training.

10. **Never** remove fuses from high voltage electrical circuits with your bare hands. Use tongs or fuse pullers to prevent electrocution.

11. **Never** enter a boiler or pressure vessel unless a reliable person is stationed nearby to note any personal mishap or weakness.

12. **Never** enter a pressure vessel that contained a toxic agent or may contain dead air without a thorough purging. If necessary, wash and clean the surfaces prior to entering.

13. **Never** enter the furnace of an automatically-fired boiler without verifying that the firing mechanism is in the locked out position and is tagged to that effect.

14. **Always** watch for falling slag when entering the furnace of a watertube boiler.

15. **Never** enter a boiler installed in battery without securely closing and tagging all stop valves on the connecting pressure lines. If bleed-off connections occur between the stop valves, open them.

16. **Always** vent a boiler prior to draining to prevent vacuum formation that can result in a collapsed shell or tubes.

17. **Never** remove a handhole or manhole cover plate from a vessel or boiler unless it is empty.

18. **Never** use a ladder without checking for broken or worn rungs and loose side rails.

19. **Never** place a ladder on or straddling electric wiring.

20. **Never** use a ladder that is too short for the job. In trying to reach up, the steep incline of the ladder may cause you to fall backwards.

21. **Always** extend your ladder above the opening of an inspection port when entering. If you try to climb from the uppermost rung, the ladder can tip out at the top or kick out at the bottom.

22. **Always** use a ladder properly suited for the job. The bottom ends of the rails should be fitted with spikes or cleats to

prevent slipping. When positioning a ladder on a metal floor, secure the rail bottoms to a fixed object.

23. **Never** rely on someone to steady your ladder.

24. **Never** use ladders as bridges.

25. **Never** attempt to make an inspection from flexible equipment such as a rope ladder or bosun's chair unless physically capable.

26. **Always** inspect the rope when using a rope ladder or bosun's chair.

27. **Never** inspect a boiler or pressure vessel while it is under hydrostatic pressure test without taking proper precautions.

28. **Always** vent a vessel before a hydrostatic test is initiated.

29. **Never** operate soot blowers unless the burners are operating at a high firing rate.

30. **Always** adjust air flow so the flue gases contain a high carbon dioxide but low oxygen content.

31. **Never** enter a building or other enclosure in the presence of ammonia fumes, sewer gas, natural gas or any other toxic gas without proper clothing and respiratory equipment, and then only with extreme caution.

32. **Never** stand in front of the discharge opening of a relief valve, particularly during testing.

33. **Never** blowdown an appliance on an boiler without observing its point of discharge.

34. **Never** blowdown a boiler under pressure by opening the slow-opening valve first. **Always** open the cock first and then the slow-opening valve. After blowdown, close the slow-opening valve first. This prevents a sudden back pressure wave on the connecting piping.

35. **Never** shut a valve off suddenly in a pressure line except for an emergency situation. The sudden change in flow may result in rupture at a weakened location.

36. **Never** open a valve suddenly (particularly in steam line). Water hammer may result.

37. **Never** operate a pressure vessel with only part of the head or cover plates, bolts, lugs or clamps in place.

38. **Never** continue to use any bolt, pin, lug or clamp after it becomes worn, sprung, stripped or otherwise weakened.

39. **Never** release the holding mechanism of any quick-opening door or end closure until the chamber is void of pressure.

40. **Always** test automatic controls and keep them clean and regularly maintained.

41. **Never** operate boilers, pressure vessels or machines at pressures or speeds above rating.

42. **Never** depend on automatic controls as safety devices.

43. **Always** perform your own tests especially—on automatic controls.

44. **Never** neglect your fire protection equipment.

45. **Always** check fire buckets and extinguishers on a regular basis.

46. **Never** leave a water line valve connected to a sprinkler system shut off. Red tag it and stay with it until corrections or alterations have been completed; then return all valves to their proper positions for service.

FIRE PREVENTION

No one can afford to be complacent about fire and its prevention, especially in boiler rooms. An operator is often assigned to a particular fire station in the event a fire breaks out. The operator must know the location of alarms and smoke detectors within his designated work area, as well as those in other sectors. When a fire alarm is activated, this knowledge allows an operator to determine where a fire has broken out and, when appropriate, which exit to use. Knowing the location of the main panel board and how to read it is imperative in securing units to help curtail potentially hazardous situations.

Three elements must be present to start a fire:

- Fuel or volatile substances
- Adequate heat (ignites combustibles)
- Air supply (oxygen for combustion)

To extinguish a fire, one or a combination of the following must be accomplished: cut-off fuel supply, cool the burning material below its ignition point, cut off the air supply or smother the fire with sand, dirt or appropriate fire retardant chemicals.

Still, before attempting to extinguish a fire, it is essential for an operator to know what type of fire has broken out.

Fires: Fires are distinguished as follows:

- **Class A:** Fires fueled by wood, paper, coal and common materials found in trash. These are the most common fires. Once a Class A fire has progressed past the smoldering stage and the material begins to burn, it often escalates quickly and generates a tremendous quantity of heat.

- **Class B:** Fires fueled by grease, oil, and flammable liquids such as gasoline, paint and similar materials that create gases when heated. These fires typically involve petroleum products, liquid solvents, cleaners or paints that have vaporized. Recall from the combustion discussion that as fuels evaporate a volatile gas is produced which explodes and burns. Some vaporized liquids produce invisible, heavier than air fumes which flow along the floor, until they are ignited. Lighter than air gases collect at ceiling level. As they accumulate and become more dense, the gases can explode with sufficient force to blow out brick walls. Consequently, only use flammable liquids outside or in thoroughly ventilated areas.

- **Class C:** Fires caused by shorts in electrical wires, motors, switches, fuse boxes and appliance circuitry. Class C fires result from overheating, due to circuit overloads, and broken or frayed wire insulation. In the event of an electrical fire, attempt to disconnect the component which caused the fire. Failing that, turn the power off at the fuse box or circuit-breaker.

Note: Never try extinguishing an electrical fire with water. The electrical current can travel along a line of moisture and cause electrocution. Electricity can also travel through a metal cabinet or container and cause any combustible materials stored within to burst into flames.

Fire Extinguishers: The most common fire extinguishers contain either water (soda acid), foam or carbon dioxide. Each is used to extinguish a particular type of fire.

An operator must understand the capabilities and limitations of fire extinguishers if they are to be used properly. If a fire breaks out, rarely will an operator have sufficient time to read the label on a extinguisher to discern what type of fire it can be used on. To simplify the process so the appropriate extinguisher can be selected at a glance, both colors and geometric symbols are employed to differentiate between types of extinguishers.

Extinguishers used for Class A fires are designated with a triangle, Class B with a square, Class C with a circle. Some fire prevention equipment manufacturers also color code the equipment, green represents Class A, red represents Class B, and blue represents Class C. Often, pictures accompany the geometric symbols to show which type of fire can be put out with the extinguisher.

- **Class A Extinguishers**: These are commonly used on fires fueled by wood, paper, plastics and rags and usually contain either water or a weak solution of sodium bicarbonate (baking soda). Not equipped with valves, safety mechanisms or triggers, these extinguishers only require to be turned upside down to operate. When overturned, a small vial of acid inside the tank mixes with the soda water solution producing a gas that displaces the water in the tank and forces it out the hose. Some extinguishers are designed with handles that must be pumped to expel water. Even buckets of water are considered Class A extinguishers. It is imperative that Class A fires be drenched with water after the initial blaze has been contained because the core of a wood or rag fire, for example, may still be active even though the surface fire has been extinguished.

- **Class B Extinguishers**: The most effective extinguisher in this category contains a dry chemical that is expelled in as a powder to smother a fire. The use of foam extinguishers is limited to small, contained fires for the most part. Some extinguishers use carbon dioxide as a fire retardant, but this material leaves a residue which must be cleaned up after a fire has been put out.

- **Class C Extinguishers**: These pressurized extinguishers utilize dry, nonconductive chemicals to smother fires.

- **ABC Extinguishers**: These extinguishers are labeled with multiple designations because they can effectively extinguish various classes of fires. This "all purpose" extinguisher is advantageous because personnel are not required to determine which type of fire they are fighting. Since most ABC extinguishers are under pressure, applications are made from a safe distance of eight to ten feet.

In addition to alphabetical designations, some extinguishers also have numerical ratings that indicate their effectiveness against various fires. For instance, an extinguisher labeled 1A:10BC indicates its effective against a moderate fire fueled by paper, grease or an ignited liquid such as degreasing solvent. A extinguisher labeled 2A:10BC is adequate for putting out a fire twice the size as the one in the preceding example. (Note that Class C extinguishers are effective and safe for use on electrical fires. These carry no numerical designations.)

Remember, even with small or insignificant fires, never take time to fight a fire if you risk having an exit blocked. Despite the fact that you may have the proper equipment on hand, fire fighting takes time. If a fire cannot be sufficiently contained, and you have tried to extinguish it, valuable time has been lost. Fires can cut off escape routes in a matter of seconds. If any doubt exits about whether a fire can be totally extinguished, no attempt should be made to fight it. Instead, sound an alarm, leave the endangered area and contact the fire department.

When vacating an area where fire has broken out, never attempt to use and elevator to escape. During a fire situation, if a building contains several elevators, they are programmed to immediately return to the main floor and remain there unless operated by a special elevator key. Most buildings with elevators usually have one elevator that serves as an access for firemen and their equipment.

CONCLUSION

Keep all work areas clean and free of debris to minimize the hazard of fire. All combustible liquids should be stored in spill-proof safety containers. Always wipe up water and oil spills at once to prevent accidents. Do not use gas, naptha or carbon tetrachloride for cleaning. Store oily rags in approved containers.

Check your work areas for any unsafe conditions and correct them. For instance, cap low lying or protruding OS&Y valve stems with a rubber end to protect personnel.

To prevent accidents and permanent or temporary disabilities to personnel, **work safely**. Remember, good housekeeping and a clean, well marked boiler room makes a safe work place.

APPENDIX B
BOILER OPERATOR'S PREPARATORY LICENSING EXAMINATION

Multiple Choice:

1. 1st class license should have at least:
   a. (1) (2) (3) (4) year(s) experience in operation or assisting in operation of 1st class equipment.
   b. (1) (2) (3) (4) year(s) experience in operation or assisting in operation of boiler over 15 psig steam or psig water.
   c. (1) (2) (3) (4) year(s) experience in operation of or assisting in operation of refrigeration on AC unit exceeding 100 tons.

2. 1st class limited should have at least:
   a. (1) (2) (3) (4) year(s) experience in operation or assisting in operation of 1st class limited equipment.
   b. (1) (2) (3) (4) year(s) experience in operation or assisting in operation of low or high pressure boilers.

c. (1) (2) (3) (4) year(s) experience in operation or assisting in operation of refrigeration or AC units exceeding 100 tons.

3. Special license applicants shall have at least:
(1) (2) (3) (4) year(s) experience in operation of equipment of the type covered by such license.

4. 3rd class all applicants shall have at least:
(1) (2) (3) (4) year(s) experience in operation or assisting in operation of low or high pressure boilers.

5. The following licenses must have a notarized letter from former employees stating work experience.
a) 1st b) 1st limited c) Special d) 3rd

6. Re-examination. A new exam shall be held at intervals of not less than
a) 3 months b) 4 months c) 1 month d) 2 months

7. Re-examination fee is:
a) $5.00 b) $15.00 c) $20.00 d) $25.00

8. Renewal license fee is:
a) $5.00 b) $15.00 c) $20.00 d) $25.00

9. Duplicate license fee is:
a) $5.00 b) $15.00 c) $20.00 d) $25.00

10. A certificate will be issued with a grade of:
a) 65 b) 70 c) 75 d) 80

11. Vent:
    a) air removed from a space and not reused
    b) a tube or conduit used for conveying air
    c) a conduit or passageway, vertical or nearly so, for conveying product of combustion to the outside atmosphere
    d) a duct which extends vertically one full story or more

12. A device having a predetermined temperature fusible member for the relief of pressure
    a) safety valve
    b) fusible plug
    c) fusible member
    d) pressure relief device

13. Baffle:
    a) a device used for the transfer of heat from one medium to another
    b) Ed LaRocque course on stationary boiler operation
    c) a device to prevent sparks, embers or other ignited material above a given size from being expelled into to the atmosphere
    d) any construction intended to change the direction of flow of products of combustion

14. What is the most prominent cause of furnace explosions?
    a) repeated attempts to light-off without purging
    b) ignition of an accumulated combustible mixture within the boiler furnace enclosure
    c) accidental ignition due to faulty wiring
    d) accidental ignition due to flames from welding

15. Which of the following is not a common cause of furnace explosions?
    a) failure of ignition energy to burners
    b) fuel leakage into an idle furnace
    c) collection of sewer gases in the furnace
    d) repeated unsuccessful attempts to light-off without purging

16. Which of the following might be the indirect cause of an oil fired furnace explosion?
    a) using a low viscosity oil
    b) the incompressibility of oil
    c) collection of sewer gases in the furnace
    d) repeated unsuccessful attempts to light-off without purging

17. With alternate fuel firing, which of the following is required?
    a) complete shutdown prior to switch-over
    b) an automatically positioned fuel selector switch
    c) a manually positioned fuel selector switch
    d) none of the above

18. What condition might result from having fuel oil overheated when it arrives at the burner?
    a) it may create a vapor lock
    b) it will lose part of its heating capacity
    c) it might vaporize and cause an explosion
    d) it might solidify and block the burner

19. The flash point of #6 fuel oil is approximately:
    a) 100°F
    b) 150°F
    c) 200°F
    d) 300°F

20. The permissible percent of water and sediment in #6 fuel oil is ______ by volume.
 a) trace
 b) 0.50
 c) 1.00
 d) 2.00

21. Steam containing some water at its saturation temperature is:
 a) superated steam
 b) dry saturated steam
 c) wet saturated steam
 d) ambient steam

22. Even small leaks of fuel oil can create potential fire hazards because:
 a) fuel oils have high volumetric heats of combustion
 b) oil is more volatile than gas
 c) spontaneous combustion is more likely to occur with oil
 d) oil fires cannot be extinguished with foam

23. When a boiler is to be operated at a pressure greater than 400 psi, not less than ______ glasses must be provided.
 a) one
 b) two
 c) three
 d) four

24. For power boilers having a heating surface of not more than 100 ft^2, the feedwater pipe shall not be less than:
 a) ½″ pipe size
 b) 1″ pipe size
 c) two ⅜″ pipes
 d) two ½″ pipes

25. A conventional, float operated low water fuel cut-off will turn off ______ yet will not mechanically supply feedwater to a boiler.
   a) a thermostatic fuel burner
   b) an automatic fuel burner
   c) an automatic water feeder
   d) an electric water feeder

26. When a globe valve is used in the feedwater line:
   a) the valve must be installed so that the boiler pressure is beneath the valve disc so the valve closes against feed pressure
   b) the valve must be installed so that the feedwater pressure is beneath the valve disc so the valve closes against feed pressure
   c) the valve may be installed in either of the above methods
   d) globe valves cannot be used in feedwater lines

27. The feed pipe shall be provided with:
   a) two check valves, one near the boiler
   b) a check valve near the boiler and a valve or cock between the boiler and the check valve
   c) two check valves with a globe valve between them

28. Which of the following gasket materials is not limited as to pressure?
   a) hard rubber
   b) neoprene
   c) metal asbestos
   d) cement asbestos

29. What is required between the pressure gage and the boiler?
    a) a trap or a siphon
    b) a vacuum breaker check valve
    c) cross fittings to facilitate cleaning
    d) a drain cock

30. The water and steam connections of a water column to a power boiler shall not be less than:
    a) ½″
    b) ¾″
    c) 1″
    d) 1½″

31. Saturated steam at a pressure of 150 psig passes through a pressure reducing valve, reducing the pressure to 15 psig. What is the temperature of the leaving steam?
    a) 366°F
    b) 212°F
    c) 100°C
    d) 250°F

True and False:

1. A 1st Class Engineer's License qualifies one:
   a) ______ to work on all steam generating boilers
   b) ______ superheaters
   c) ______ pressure vessels
   d) ______ refrigeration
   e) ______ A/C of limited tonnage

2. A 1st Class Limited License qualifies one to operate and maintain:
   a) ______ all boilers subject to a pressure exceeding 15 psig
   b) ______ 30 psig water boilers
   c) ______ refrigeration and A/C of unlimited tonnage

3. A 3rd Class Licensee may operate and maintain:
   a) ______ heating boilers not to exceed 20 psig steam
   b) ______ 30 pounds of water safe working pressure

4. A Steam Special Licensee may operate and maintain:
   ______ all boilers, engines, pumps and apparatus(es) of unlimited size and pressure

5. A Steam Special Limited Licensee may operate and maintain all boilers, engines, pumps and apparatuses in conjunction with laundry and dry cleaning process equipment only:
   ______ not to exceed 40-HP subject to a pressure not to exceed 150 pounds

6. Mechanical equipment that requires licensed operators include:
   a) ______ all steam boilers of 10-HP or more, subject to an operating pressure of more than 15 lbs/in^2
   b) ______ all steam boilers subject to an operating pressure of 15 lbs/in^2 or less and all hot water boilers with a temperature exceeding 212°F

7. Daily operator log required by all operators:
   ______ showing time he arrives and initials in full view with certificate and license

8. ______ A furnace is an enclosure for the combustion of fuel.

9. ______ A fuel rich mixture is a ratio of air to fuel which provides more air than required for an optimum air/fuel ratio.

10. ______ A recycle is a start-up initiated by steam pressure following a normal shut down.

11. ______ An explosive mixture is a flammable mixture in a confined space.

12. ______ To light-off is to establish combustion of fuel entering a furnace.

13. ______ A low water cut-off is a device arranged to indicate that the boiler system water is at a dangerously low level.

14. ______ A re-start is a mechanically initiated start-up.

15. ______ A set point is a predetermined value to which an instrument is adjusted and at which it shall perform its intended function.

16. ______ The outlet draft is the flue gas pressure at the outlet of the flue.

17. ______ Recycle is a start-up initiated by fuel pressure following a normal shut down.

18. ______ To monitor is to sense and alarm a condition requiring attention, and initiate corrective action.

19. ______ The operating range shall be determined by test.

20. ______ A soot blower is a mechanical device for introducing steam or air to clean heat absorbing surfaces.

21. ______ An alarm is an audible or visual signal indicating a dangerous condition.

22. ______ Air change volume must be calculated at 14.7 psig and 70°F.

23. ______ To supervise is the same as to monitor.

24. ______ Boiler feedwater treatment is generally not necessary in your area.

25. ______ The reset button on an oil burner is to adjust the rate of oil flow.

26. ______ Feedwater should be heated to approximately the same temperature as the system water before being fed to the boiler.

27. ______ Feedwater should enter the boiler at a point above the normal water level.

28. ______ An economizer is a device which gives greater burning efficiency to the oil.

29. ______ A gate valve causes a greater pressure loss than a globe valve.

30. ______ A vapor seal is required on the outside of insulation covering steam lines.

31. ______ A licensed Steam Generating Boiler and Boiler Piping Master is permitted to do any type of welding connected with his trade.

32. ______ The temperature of the flame in a boiler can be measured with a pyranometer.

33. ______ A single return may be used to serve both high and low pressure steam systems.

34. ______ A Hartford Loop is used to prevent water from siphoning from the boiler.

35. ______ A dry return carries no condensate.

Fill in the blanks:

1. Stop valves shall be located in the supply and return of boiler piping to facilitate ____________.

2. Access openings to a furnace located in an attic shall be no more than ____________ feet from the furnace.

3. The word "shall" in the BOCA code means ____________.

4. Gas piping must be supported every ____________ feet.

5. Maximum length or vibration isolation connectors per duct is ____________ inches.

6. The ratio on net heat removal to the rate of total energy input expressed in consistent units and under designated rating conditions is ____________ performance.

7. A gas tight joint obtained by the joining of metal parts and metallic alloys used in a system with operating temperature from 400 to 800°F is a ____________ joint.

8. Application for permit shall be deemed abandoned ____________ months after date of filing if permit has not been issued.

9. The ____________ is not responsible for testing of new installations when required—including providing test equipment.

10. What is the minimum test period for piping in a radiant heating system?

11. Before gas appliances are connected, piping systems shall stand a pressure of at least ____________.

12. When a first class operator is delinquent on renewing his license for more than ____________ year(s), he will be required to take the examination again.

13. Where shall an operators license be posted?

14. When performing a hydrostatic test on a boiler from the factory, the pressure should be raised to ____________ times operating pressure.

15. To qualify for a First Class License you must have a minimum of ____________ year(s) experience on high pressure boilors.

16. Under what boiler conditions will an alarm sound?

17. On a feedwater supply line, what type of valve is properly used?

18. When a pressure relief valve is leaking at a pressure lower than its rating, what should be done?

19. What chemicals should be used to control algae growth in a cooling tower?

20. A second safety valve should be installed on a boiler that has more than ____________ heating surface.

21. On a boiler, where should the safety relief valve be located?

22. What type of valve is used on a low water cut-off supply line?

23. For what reason can an operator not be fired?

24. When exicuting a blowdown, what valve is opened last and closed first?

25. What is an operator's first task or responsibility when taking over a shift?

26. When cleaning heating surfaces on a boiler, what method should be used?

27. ______________ gallons = ft^3

28. ______________ cubic inches = 1 gallon

29. ______________ pound(s) of water = 1 gallon

30. ______________ ° Celsius = 100° Fahrenheit

31. ______________ pounds absolute = 14.7 psig

32. What type of fire extinguisher would you use on an oil fire? Electric fire?

APPENDIX C
SOLUTIONS REVIEW QUESTIONS

SECTIONS 1–6

1-1. hydronics

1-2. has greater specific heat than other fluids and carries a greater amount of heat/pound

1-3. circulating pump

1-4. a) yes b) for combustion

1-5. aquastat

1-6. balance pressure of system

1-8. a) pressure relief valve
b) relieves pressure by discharging water

1-9. heating elements immersed in water

1-10. strongly constructed, closed vessel filled with water

1-11. gas, oil, coal

1-12. heat content of a substance

1-13. 212°F

1-14. apply heat to a larger surface area

1-15. allow air/fuel to mix in greater quantities

1-16. more water comes in contact with heated metal—faster steam production

1-17. replenish system water
1-18. heat exchanger—condenses steam
1-20. device doesn't allow steam to return to feedwater tank
1-21. drawn to tank by vacuum pump
1-22. oil, gas, coal
1-23. natural, bottled gas
1-24. abundant, needs no preparation for burning, requires no storage, burns cleanly, economical
1-25. regulator
1-26. city gas line pressures may vary
1-27. natural or mechanical draft
1-28. fire box or combustion chamber
1-29. pilot burner
1-30. suction pump
1-32. back to tank via return line
1-33. anthracite (hard coal), bituminous (soft coal)
1-34. feeds coal into fire box
1-35. screw feed, ram feed
1-36. natural rising process of hot air due to difference between atmospheric pressure and lower pressure existing inside boiler
1-37. provides combustion air, vents combustion by-products (flue gases)
1-38. natural, mechanical
1-39. increasing temperature of flue gases increases draft velocity
1-40. dampers, regulators, chimney height
1-41. a) mechanical draft (forced or induced)
b) burns inferior grades of fuel, maintains constant steam pressure as load demands change
1-42. radiation, convection, conduction
1-43. radiation
1-44. heat transfer without particle displacement
1-45. convection

1-46. a) yes b) fireside where flames of combustion contact metal plates
1-47. amount of heating surfaces and the arrangement of distribution
1-48. deposits (soot) insulate and reduce heat transfer

2-1. A boiler is a closed vessel that uses combustible fuels or or electricity to heat water and convert it into steam.
2-3. The side of the boiler surface exposed to the heat of combustion determines the size of the heating surface. Of the two boilers with identical diameters, the water-tube boiler has a larger heating surface than a firetube boiler.
2-4. a) The Scotch Marine boiler frequently has more than a single furnace.
b) Multiple furnaces generate more heat in a given period of time than can a single furnace. The additional heat produced makes the SM boiler a fast steamer.
2-5. a) No. b) Fire tubes cannot burst because they are under the influence of external pressure. Consequently, they can only collapse.
2-6. a) The watertube boiler is safer. b) If a watertube ruptures within the boiler, only a small quantity of water is released and will instantly vaporize or flash to steam.
2-7. vertical tube boiler
2-8. dry top has water line below the upper tube sheet, wet top has water line above the upper tube sheet
2-9. locomotive boiler
2-10. A steam dome helps produce drier steam and also slightly increases the steam storage capacity of a boiler
2-11. cast iron (ci) boilers

2-12. The capacity of cross drum boilers can be increased by lengthening the drum to accommodate attachment of additional tubes. The capacity of long drum boilers can only be increased by installing additional drums to the boiler.

2-13. A, D, O

2-14. Additional drums allow increased water storage and steam capacity. Multiple drums also help meet demand fluctuations while minimizing pressure drop.

2-15. equalizing tubes connect the steam drums to help equalize steam pressure in the drum series

2-16. tubes are bent to facilitate expansion and to ease tube replacement

2-17. a) sediment and boiler impurities settle out and collect in the mud drum b) once settled, the mud (sludge) can be blown down or removed from the boiler to prevent it from building up and restricting water passages inside the boiler

2-18. a) dry-back, wet-back

2-19. a design that promotes constant water circulation for even heating; a large heating surface for quick heat transfer; ample steam storage so steam can separate from the water for dry steam production; suitable furnace space to allow complete and efficient combustion; strong and simple construction to ease accessibility for maintenance and repair is essential.

3-1. check valve; feedwater line pressure

3-2. vacuum pump

3-3. a) where ever quick, positive shut-off is required
b) offers a restricted water passage

3-4. cut-offs: float magnet, float linkage, electrode probe
3-5. a) feedwater regulator
b) electronic sensor activates feedwater pump
3-6. false
3-7. simulate an actual low water condition
3-8. under pressure and firing
3-9. a) shut-down and cool boiler
b) replenish water supply
c) immediately replace low water cut-off
3-10. vacuum created by air pump
3-11. continuous, float only, float or vacuum
3-12. rising water or decreased vacuum actuates vacuum pump switch
3-13. a) steam gage
b) highest point of steam space
3-14. siphon
3-15. compared to inspectors gage attached at T fitting
3-16. a) under 400°F, 1.5 MAWP of safety valve
b) over 400°F, 2 times set pressure of safety valve
3-17. bourdon tube, diaphragm
3-18. indicates water and steam level
3-19. 3 feet above highest heat surface
3-20. water column; diverts water to gage glass and arrest water level surges
3-21. plain try cock—wheel handle
sprint try cock—pull chain
3-23. a) air cock
b) prevents pressure build-up and vacuum
c) highest drum of a boiler
3-24. fusible plug; brass fitting with tin core that drops out when low water condition occurs
3-25. a) pop safety valve
b) highest point of steam side of boiler—away from walkways

3-26. a) maximum allowable working pressure
b) low pressure up to 15 lbs psig; high pressure exceeding 15 lbs psig

3-27. quick opening and closing of safety valve without sticking or hesitating—indicates malfunctioning or steady safety valve

3-28. a) if heating surface exceeds an area of 500 ft^2
b) 6% over MAWP

3-30. increase boiler pressure to pop valve or manually test by opening valve lever

3-31. a) minimum of once each week
b) sticking and blockage due to corrosion

3-33. a) blowdown valve b) screw type gate valve
c) 25% greater than valve's MAWP

3-34. a) lever operated gate valve
b) provides positive seal and absorbs majority of wear

3-35. a) location b) surface and bottom blowdown surface—controls foaming; bottom—high water control, sludge and sediment removal; chemical concentration control, dumping

3-36. a) allows sediment build-up
b) where subject to little operation

3-37. mud drum

3-38. gate valves; unobstructed flow, positive seal, opens and closes slowly, doesn't trap sediment

3-39. rising stem valve indicates whether it's open or closed at a glance

3-40. automatic non-return valve; promotes steam flow from boiler to main and checks backflow

3-41. cannot be partially closed—throttling results in wear of seating surfaces preventing positive shut-off

3-42. a) controls throttling without excessive wear
b) Z flow pattern eliminates chattering
c) easily serviced, can be repaired

3-43. a) steam trap
b) drains off condensate from steam system and prevents water hammer
3-44. water hammer results from priming or carry-over
3-45. thermostatic, mechanical, thermodynamic
3-46. return traps route condensate back to boiler; non-return discharge condensate as waste
3-47. inverted bucket trap; flexible, simple internal construction, can be modified in field, wide array of sizes and capacities, temperature has no effect
3-48. a) used with convectors or radiators in space heating systems
b) balance-pressure and float-thermostatic traps
3-49. tilting and floating traps
3-50. waterbound radiators/convectors, condensate flashes back into steam, vacuum pump becomes steambound, boiler overheats
3-51. a) temperature readings b) strap on thermometers; temperature sticks

4-1. oxidation process/chemical recombination
4-2. air, fuel, heat for ignition
4-3. a) incomplete combustion lacks adequate oxygen supply to completely consume fuel
b) secondary air
4-4. a) incomplete: carbon, soot, aldehydes, carbon monoxide
b) complete: water, carbon dioxide, heat
4-5. a) greater air to fuel ratio (10-30% excess air)
b) reduction of operating efficiency and furnace temperature

4-6. primary air initiates combustion and controls combustion rate

4-7. maximize efficiency and minimize air pollution

4-8. coal, gas, oil

4-9. a) hydrocarbons b) natural gas

4-10. a) non-luminous b) more chemical energy is transformed into heat

4-11. a) flash point: temperature where volatile elements flash when exposed to open flame
b) fire point: temperature that sustains continuous burn

4-12. steam pressure changes

4-13. on-off control, positioning response and metering control

4-14. locks out power so boilers can't fire

4-15. a) metering control b) measures air and fuel flows, also draft loss to compensate for variables

4-16. purge, lights pilot, lights main burner

4-17. a) fire-eye b) pilot or flame failure (lack of light) shuts off fuel supply and turns off burner

4-18. detects light, conducts electricity to complete current

4-19. ability of fuel to be converted into gas so it's combustible

4-20. a) oil: atomizing b) coal: pulverizing

4-21. to expose maximum surface area; to facilitate mixing with primary air so can vaporize and burn

4-22. mixes oxygen (secondary air) with burning fuel for complete combustion

4-23. high hydrocarbon contents

4-24. a) hydrogen b) higher hydrogen concentration, quicker fuel burns

4-25. zero balance reducer or governor

4-26. vaporstat; gas line pressure

4-27. diverts gas to pilot

4-28. size of gas orifice and manifold pressure

4-29. aids in mixing air/fuel

4-30. suction created by fuel flow pulls air and fuel into mixing chamber (mixjector)
4-31. draft
4-32. due to increased line pressure
4-33. a) venturi tube b) no
4-34. pilot flame
4-35. a) flame rod or thermocouple b) sense heat and opens electrical current
4-36. see lighting sequence
4-37. resistance to flow, indicates oil weight
4-38. higher viscosity, higher Btu content
4-39. a) 2, 4, 5, 6 b) 6
4-40. a) cold oil resists pumping, clogs nozzles, results in poor atomization
b) electric and steam heaters
4-41. atomizing and vaporizing b) broken into droplets by steam pressure, forced through a nozzle, centrifugal force applied
4-43. equipped to burn various fuels, most commonly gas and coal
4-44. a) atomize oil and deliver fixed amount of fuel to combustion chamber b) central component of main burner
4-45. clogging or fouling of strainer baskets
4-46. a) anthracite, bituminous, lignite b) anthracite
4-47. soft coal
4-48. a) hand, mechanical stoking, pulverized firing
b) pulverizing
4-49. spreading, coking, alternate
4-50. alternate sides; front-to-rear
4-51. a) sprinkler, chain grate overfeed, underfeed
b) reduction of labor, precise firing
4-52. ram feed; screw feed

5-1. a) check water level in on-line boilers
b) blowdown gage glass

5-2. to discern what occurred so precautions can be taken should a hazardous condition exist or have the potential of developing

5-4. blocking it with hand

5-5. light, bright color with split ends

5-6. boiler frequently takes on water

5-8. vacuum

5-10. remove sludge, control high water or chemical concentration or to dump

5-12. annually

5-13. so sludge accumulations do not harden

5-14. if the boiler was inspected recently

5-15. a) leak test b) 1.5 time MAWP

5-16. sweating can result and be mistaken for a leak

5-17. a) lay-up b) protect against corrosion (oxidizing agents)

5-18. a) dry lay-up b) quick lime, silica gel c) every month

5-19. glass expansion

5-20. washers

5-21. 2-3 inches above highest heating surface

5-22. normal operating water level

5-23. some water is visible in the gage glass or when water comes out of the bottom gage glass try cock

5-24. carry-over and water hammer

5-25. failure to purge combustible gases

5-26. human error

5-27. soap bubble test

5-28. oil, gas

5-29. purge furnace box before igniting

5-30. turn the clock ahead to by-pass the purge cycle

6-1. scale, corrosion, fouling, foaming
6-2. solvent
6-3. dissolved mineral salts, dissolved gases, suspended solids
6-4. mineral salts
6-5. a) scale b) as water is added to boiler salt concentration increases until water becomes saturated and salts precipitate out
6-6. insulate, corrodes
6-7. watertubes: impaired circulation, higher operating pressure
6-8. a) coagulated mineral salts or "mud"
b) blowdown of mud drum
6-9. a) deterioration of metal surfaces
b) pitting, grooving, cracking, galvanic & uniform corrosion
6-10. rust or oxidation
6-11. soot, sulfur gas from incomplete combustion
6-12. a) foaming b) high concentration of impurities
c) surface blowdown, anti-foaming chemicals
6-13. filtration
6-14. sand, silt, grease, oil, bacteria, algae
6-15. a) water's tendency to corrode metal or form scale
b) 7.0
6-16. parts per million
6-17. ability of water to neutralize acid
6-18. calcium carbonate
6-19. amount of dissolved calcium or magnesium salts
6-20. chlorides
6-21. concentration of an element in system water that risen above concentration in makeup water
6-22. solids
6-23. blowdown dilutes concentration

6-24. chemical that coats metal surface with protective film and prevents electrochemical corrosion
6-25. feedwater heater, lines: internal steam drum surfaces
6-26. a) higher temperature—faster corrosion rate
b) extremely fast circulation rate—higher corrosion rate; extremely slow circulation rate—higher corrosion rate
6-27. prevent corrosion; promotes scale formation
6-28. scale inhibitor
6-29. soda ash, caustic soda, sodium silicate, sodium phosphate
6-30. a) sulphite b) non-harmful sulphate
6-31. oxygen scavengers (sodium sulfate, hydrazine)
6-32. no standard water treatment exists
6-33. a) carry-over: transference of water impurities to steam system b) priming: transference of water into steam lines
6-34. steam bubbles become encased by a film of water—foaming
6-35. reduced impurities (controls foaming), reduced heat and chemical losses
6-36. zeolite water softeners, deaerating heaters, evaporators, open or closed feedwater heaters
6-37. brine solution of sodium chloride
6-38. removes non-condensible gases from feedwater
6-39. oxygen less soluble as water temperature rises
6-41. shut-down boiler; severe oil induced foaming
6-42. a) boil out with alkaline material; boiler may need to be retubed b) oil cannot be removed from tube surfaces
6-43. a) no b) may be contaminated with oil

PREPARATORY EXAMINATION ANSWERS

Multiple Choice

1. a) 3
   b) 3
   c) 3
2. a) 3
   b) 1
   c) 3
3. 3
4. 1
5. d
6. c
7. d
8. a
9. a
10. c
11. c
12. b
13. d
14. b
15. c
16. b
17. d
18. a
19. b
20. d
21. b
22. a
23. a
24. a
25. b
26. b
27. b
28. c
29. a
30. a
31. d

True and False

1. a) T
   b) T
   c) T
   d) T
   e) F
2. a) F
   b) T
   c) T
3. a) F
   b) T
4. T
5. F
6. a) F
   b) T
7. T
8. F
9. T
10. T
11. T
12. T
13. F
14. F
15. T
16. F
17. F
18. F
19. T
20. T
21. F
22. T
23. F
24. F
25. F
26. T
27. F
28. F
29. T
30. F
31. T
32. F
33. T
34. F
35. T

Fill in the blank

1. Drain boiler without draining the system.
2. 120
3. mandatory
4. 10
5. 10
6. coefficient of
7. soldered
8. 6
9. contractor
10. 4 consecutive hours
11. not less than 3 psi.
12. one
13. Mechanical room office wall
14. Not exceeding 50% maximum working pressure
15. 3
16. To bring a specific condition to the attention of the operator or any abnormal conditions
17. globe
18. operator should lift try-lever in an attempt to reset
19. Chlorine, copper, sulphite
20. 500-ft^2
21. top and side of boiler
22. a flushing valve
23. poor attitude
24. valve closest to boiler
25. check water levels, pressure noises
26. thoroughly sweep all walls, tubes and drums
27. 7.48
28. 2.31
29. 8.33
30. 37.8
31. 29.4
32. foam, carbon dioxide

APPENDIX D

PARTS ORDERING INSTRUCTIONS

Furnish complete information when ordering parts for your boiler. Most importantly, include the boiler serial number shown on the name plate attached to the front head of your boiler. When ordering, state the part number and the name as well as a description of the part to assure receiving the correct replacement part. Indicate when the part is required, the quantity and be certain to specify the method of shipment. If the replacement parts are for accessory equipment not shown in the operator's manual that came with the boiler, such as an electric motor pump, write down the complete name plate data from the item for which the parts are required. With this information, repair or replacement parts can then be ordered from your local manufacturer's sales representative.

When returning parts for repair, a purchase letter or an authorization letter accompanied by complete details should be sent to manufacturer's representative. Prior to returning parts, it is advisable to remove fittings or accessories from the

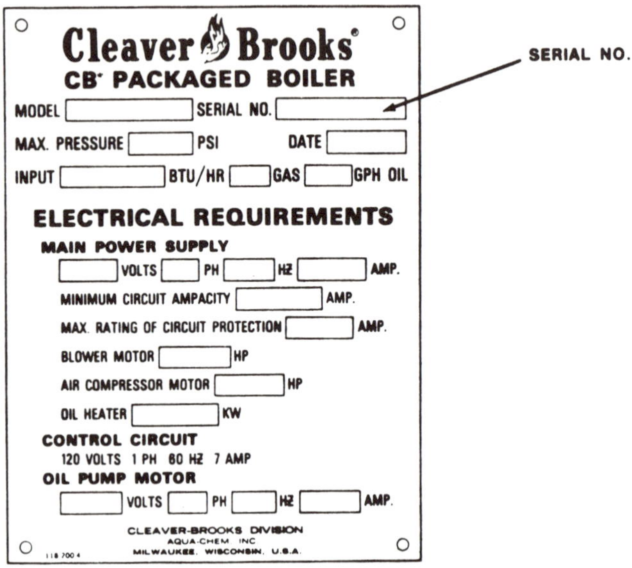

Typical name plate with serial number and other pertinent data necessary for obtaining correct replacement parts. *(Courtesy, Cleaver-Brooks)*

component to be repaired. The part must also be cleaned and drained properly to comply with shipping regulations. Always include a packing slip identifying the part and the name and return address of your company. If a part is being returned to a manufacturer for reasons other than repair or exchange, contact the representative before shipping the component. Many manufacturers require that you obtain authorization directions for the return prior to actually shipping the part.

Note that many controls and other components can be factory rebuilt or may have trade in value. These items may be available on an exchange basis, so consult with your representative.

Make The Right Connection

Need a solid working background on valves and their proper applications—this is the source book you've been waiting for!

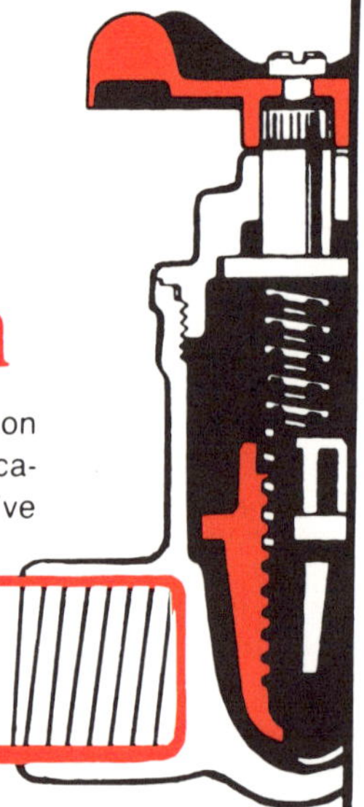

Valve Selection & Service Guide

Hard cover, 202 pages

This guide was written for contractors, servicemen and anyone who must size, select and install or repair valves on the job. The information in this book targets the air conditioning, heating and refrigeration industry, but the general principles and basic valve designs covered apply to a number of other fields as well.

Many manufacturers can help you select the right valve design required for your job, but the recommendations they make are based on the information you provide. If your information is sketchy or not thoroughly planned out, improper selections are likely to occur. The result, equipment failure, or even worse, personal injury and/or a lawsuit that can translate into considerable financial loss.

Get the Valve Selection and Service Guide so you can learn to select the proper valves for your applications and **prevent costly mistakes.** It may prove to be the best investment you'll ever make!

Business News Publishing Company
P.O. Box 2600, Troy, MI 48007

______	copies of: Boilers Simplified	$27.95 each
______	copies of: Smart Contracting Dollar & Sense Strategies	$12.95 each
______	copies of: Basic Refrigeration	$24.95 each
______	copies of: Refrigeration Licenses Unlimited	$29.95 each
______	copies of: Getting Started in Heating and Air Conditioning Service	$27.95 each
______	copies of: How to Solve Your Refrigeration and A/C Service Problems	$19.95 each
______	copies of: How To Make It In The Service Business	$19.95 each
______	copies of: The Reference Notebook Set	$15.95 each
______	copies of: The Schematic Wiring Book Set	$24.95 each
______	copies of: Hydronics, the art of cooling and heating with water	$9.95 each
______	copies of: Good Piping Practice	$5.95 each
______	copies of: The Valve Selection & Service Guide	$39.95 each
______	copies of: The Service Hot Line Handbook Vol. I	$12.95 each
______	copies of: The Service Hot Line Handbook Vol. II	$12.95 each
______	copies of: The Service Hot Line Handbook Vol. III	$12.95 each

Total number of books______ **Total $**__________
(count set as one book)

Add $2.00 shipping/handling for first book, 50¢ for each additional book.

☐Check ☐Money Order enclosed for $______

Allow 3–6 weeks for delivery.
For express delivery, add extra 15% of order total.

Send to:
Name ______________________________

Company ______________________________

Address ______________________________

City/State/Zip ______________________________

Please check one: ☐New Customer ☐Previous Customer

(Prices subject to change without notice.)